NITROGEN METABOLISM
AND THE ENVIRONMENT

NITROGEN METABOLISM AND THE ENVIRONMENT

A Joint Symposium of the American
Physiological Society and the American
Society of Zoologists held at Bloomington,
Indiana, U.S.A. in August, 1970

Edited by

J. W. CAMPBELL

*Department of Biology
Rice University, Houston
Texas, U.S.A.*

and

L. GOLDSTEIN

*Division of Biomedical Science
Brown University, Providence,
Rhode Island, U.S.A.*

1972

ACADEMIC PRESS - London and New York

ACADEMIC PRESS INC. (LONDON) LTD
24-28 Oval Road,
London, NW1

U.S. Edition published by
ACADEMIC PRESS INC.
111 Fifth Avenue,
New York, New York 10003

Copyright © 1972 by ACADEMIC PRESS INC. (LONDON) LTD

All Rights Reserved
No part of this book may be reproduced in any form by photostat, microfilm, or any other means, without written permission from the publishers.

Library of Congress Catalog Card Number: 76-187921
ISBN: 0 12 157850 X

Printed in Great Britain by
THE SCOLAR PRESS LIMITED 20 MAIN ST., MENSTON, YORKS.

CONTRIBUTORS

R. H. ADAMSON *Laboratory of Chemical Pathology, National Cancer Institute, National Institutes of Health, Bethesda, Maryland 20014, U.S.A.*

J. W. CAMPBELL *Department of Biology, Rice University, Houston, Texas 77001, U.S.A.*

J. J. CORRIGAN *Department of Life Sciences, Indiana State University, Terre Haute, Indiana 47809, U.S.A.*

R. B. DROTMAN *Department of Physiological Sciences, University of California, Davis, California 95616, U.S.A.*

R. P. FORSTER *Department of Biological Sciences, Dartmouth College, Hanover, New Hampshire 03755, U.S.A.*

L. GOLDSTEIN *Division of Biochemical Sciences, Brown University, Providence, Rhode Island 02912, U.S.A.*

A. D. GOODMAN *Division of Endocrinology and Metabolism, Department of Medicine, Albany Medical College, Albany, New York 12208, U.S.A.*

J. A. MCDONALD *Duke University Medical School, Durham, North Carolina 27706, U.S.A.*

J. MAETZ *Groupe de Biologie Marine, Department de Biologie, Commissariat a l'Energie Atomique, Station Zoologique, Villefranche/Mer 06, France.*

R. F. PITTS *Department of Physiology, Cornell University Medical College, New York, New York 10021, U.S.A.*

B. SCHMIDT-NIELSEN *Department of Biology, Case Western Reserve University, Cleveland, Ohio 44106, U.S.A.*

G. G. STEPHENS *Department of Development and Cell Biology, University of California, Irvine, California 92664, U.S.A.*

P. R. TRAMELL *Department of Pharmacology, Stanford University, Medical School, Stanford, California 94305, U.S.A.*

PREFACE

During the last decade, the number of studies dealing with the comparative biochemistry and physiology of nitrogen metabolism has grown at an accelerated pace. This has been due, in part, to the general increase in research activity in the biological sciences, but, beyond this, there is a growing awareness on the part of biologists of the importance of understanding how animals interact with their environment at the cellular and subcellular levels. The influence of environment on animal nitrogen metabolism has attracted the attention of biologists for many years. Homer Smith's pioneering work on nitrogen metabolism in estivating African lungfish focused attention to the dramatic effects that water availability has on the pattern of nitrogenous end-products in these and other animals. Joseph Needham also contributed much to this area of investigation in pointing out the influence of the embryonic environment on the pattern of nitrogen excretion in a variety of aquatic and terrestrial animals. The work and ideas of the early investigators were popularized by Ernest Baldwin in his book, *An Introduction to Comparative Biochemistry*, first published in 1936. The early generalizations were formulated when the gap between whole animal physiology, on the one hand, and test tube biochemistry, on the other, was very great. Although these early generalizations have, for the most part, withstood the test of time, certain aspects can now be reinterpreted and reevaluated in terms of modern concepts of cellular metabolism, its physical organization and metabolic control.

The contents of this volume are from a symposium entitled "Nitrogen Metabolism and the Environment" sponsored jointly by the American Physiological Society and the Division of Comparative Physiology and Biochemistry, American Society of Zoologists and held in Bloomington, Indiana in conjunction with the August, 1970 meetings of the American Institute of Biological Sciences and the American Physiological Society. In organizing this symposium, we tried to select topics representing current areas of investigation in the metabolism of nitrogenous compounds. An attempt was also made to integrate comparative aspects of the subject with the physiology and biochemistry of nitrogen metabolism in man and higher animals. We cogently omitted the topic of nitrogen metabolism in plants from the symposium since this subject is far enough removed (although not unrelated) and broad enough to require a symposium unto itself. We hope that this symposium, dealing with recent developments in the physiology and biochemistry of natural and foreign nitrogenous compounds, will help to focus attention on the cellular aspects of animal nitrogen metabolism and how it is affected by environmental factors. In addition to the participants whose manuscripts are published here, we also wish to thank Professors Mary Ellen Jones and A.W. Martin for their participation in the symposium.

         J.W.C.
         Houston, Texas
         and
         L.G.
         Providence, Rhode Island

CONTENTS

Contributors	v
Preface	vii
Nitrogen Metabolism in Terrestrial Invertebrates. J. W. CAMPBELL, R. B. DROTMAN, J. A. McDONALD and P. R. TRAMELL	1
Adaption of Urea Metabolism in Aquatic Vertebrates. L. GOLDSTEIN	55
Mechanisms of Urea Excretion by the Vertebrate Kidney. B. SCHMIDT-NIELSEN	79
Interaction of Salt and Ammonia Transport in Aquatic Organisms. J. MAETZ	105
Amino Acid Accumulation and Assimilation in Marine Organisms. G. C. STEPHENS	155
D-Amino Acid OXIDASE: Response to a Stereospecific Challenge. J. J. CORRIGAN	185
Foreign Nitrogenous Compounds—Their Effects and Metabolism. R. H. ADAMSON	209
Excretion of Foreign Nitrogenous Compounds. R. P. FORSTER	235
Control of Renal Ammonia Metabolism. R. F. PITTS	277
Relation of Carbohydrate Metabolism and Ammonia Production in the Kidney. A. D. GOODMAN	297

NITROGEN METABOLISM IN TERRESTRIAL INVERTEBRATES[1]

James W. Campbell,[2]
Robert B. Drotman,[2]
John A. McDonald,[3] and
Paul R. Tramell[4]

Department of Biology
William Marsh Rice University
Houston, Texas 77001

Among vertebrates, the evolutionary transition from an aquatic to a terrestrial environment involved, among other metabolic and physiological adaptations, a change from ammonotelism to ureotelism. Amphibians that normally undergo such a change during metamorphosis

---

[1] The experimental work was supported by grants from the USPHS (AI 05006 and 2-T1-GM-884) and the NSF (GB-8172). J.W.C. is a Career Development Awardee of the USPHS (2-K3-GM-6780).

[2] Present address: Department of Physiological Sciences, University of California, Davis, California 95616

[3] Present address: Duke University Medical School, Durham, North Carolina 27706

[4] Present address: Department of Pharmacology, Stanford University Medical School, Stanford, California 94305

of the aquatic tadpole to the semi-terrestrial adult have served as model systems for studying this transition in excretory nitrogen metabolism (6, 33, 34). The functioning of the ornithine-urea cycle to form urea for osmotic purposes in primitive marine vertebrates such as elasmobranchs (4), holocephalans (104) and possibly coelocanths (18, 93) indicates that the first "water-poor" environment in which selection favored the over-production of urea was the sea. By providing a route for the detoxication of ammonia and, of equal importance, by decreasing evaporative water loss through lowering of the vapor pressure of body fluids, the retention of the capacity to synthesize and accumulate urea by freshwater ancestors of modern amphibians allowed them to survive periods of desiccation. A functioning ornithine-urea cycle is present in lungfish, the closest extant relatives of ancestral amphibians (50) but, in the Australian lungfish, a species that is permanently aquatic and predominantly ammonotelic in its excretory nitrogen metabolism, the cycle operates to only a limited extent for excretory purposes (51). Urea accumulates in the African lungfish during estivation, a period of dehydration (73). The accumulation of urea during dehydration is of common occurrence in amphibians (7, 8, 86, 122) and also occurs in amphibians living in hypertonic environments (52). It is thus possible to envisage the general development of ureotelism in ancestral vertebrates for their survival in hypertonic and desiccating environments. Such a development prepared them for their subsequent evolutionary transition to the terrestrial environment. Once established, this mode of ammonia detoxication was utilized by all non-aquatic amphibians (29). Ureotelism continues through the "stem reptiles" (Cotylosauria) to the primitive mammals (40) to also become the predominant type of excretory nitrogen metabolism in the class Mammalia. The Reptilia are the most diversified land vertebrates utilizing either ammono-, ureo- or uricotelism (9, 61, 62). The transition to uricotelism is complete in some reptiles and this mode of nitrogen excretion was continued into the birds where it is used exclusively. The capacity for urea synthesis has been lost by birds (119).

Whereas ureotelism is a major type of excretory metabolism in semiterrestrial and terrestrial vertebrates, it is rare in the invertebrates that have invaded the land environment. Some land invertebrates utilize a modified ammonotelic type of nitrogen excretion but most are purinotelic. Purines are, for example, excreted by the most familiar terrestrial invertebrates, the arthropods. Uric acid is the major excretory product of insects although the degradation products of uric acid, allantoin, allantoic acid, etc., may also be excreted (36). The myriapodous arthropods (millipedes and centipedes) excrete uric acid (58) whereas the arachnids excrete mainly guanine (58, 102). Terrestrial crustaceans, such as the land crab *Cardisoma*, accumulate uric acid (49). Land crabs can tolerate high levels of blood ammonia and thus may excrete some nitrogen as gaseous ammonia via extrarenal routes. This latter form of "ammonotelism" has been utilized by terrestrial amphipods and isopods in their transition to the land

(57, 58, 131). Gaseous ammonia excretion can account for all of the nitrogen excreted by certain terrestrial isopods (130). Because special mechanisms, especially those concerned with the acid-base regulation of body fluids, are required to volatilize ammonia in the unprotonated form, this type of ammonotelism should be distinguished from the classical ammonotelism of aquatic species in which ammonia is released in the protonated form, NH_4^+ (114). The direct release of ammonia to the atmosphere may be more common among terrestrial invertebrates than is now known. In general, invertebrates can withstand much higher levels of ammonia in their body fluids than can vertebrates (46) and they may also tolerate wider fluctuations in pH. Both of these parameters would appear to be necessary to allow for the release of nitrogen directly as gaseous ammonia. In this context, it thus seems unlikely that ammonia toxicity has played a significant role in the adaptation of most invertebrates to the terrestrial environment. Gaseous ammonia release also occurs in certain land snails (114) but does not appear to be the major route for nitrogen excretion by these organisms (115). The land snails constitute another major group of land invertebrates that are, in general, purinotelic (25). Uric acid is the predominant excretory purine in most species, although guanine and xanthine also account for a high percentage of the excretory nitrogen. There are also recurrent reports of urea as a major excretory product in different species of land snails (25). As will be discussed below, in those species in which urea synthesis has been investigated in detail, urea appears not to be a major excretory product. Unfortunately, nitrogen excretion by slugs has not been investigated following Delaunay's report of urea as a major excretory product of the slug *Limax* (38). Earthworms are, in fact, the only common terrestrial invertebrates that have developed a fairly typical ureotelic type of excretory nitrogen metabolism (11, 12, 35). Earthworms have, however, retained their ammonotelism and appear to excrete urea only under special circumstances. Starvation is one such circumstance and certain aspects of the starvation-induced transition from ammonotelism to ureotelism in some earthworms suggest that ammonia toxicity may be a major stimulus for this change (92). Land planaria, a lesser-known but significant group of terrestrial invertebrates, have also developed a ureotelic excretory nitrogen metabolism. The cosmopolitan land-planarian *Bipalium kewense* excretes up to 73% of its total nitrogen as urea (23) and tracer studies have shown that this urea originates via the ornithine-urea cycle (21).

Evolutionary trends in excretory nitrogen metabolism among terrestrial invertebrates are not as clear as those among their vertebrate counterparts. No single invertebrate phylum has been studied in sufficient detail to allow comparisons to be made among its members similar to those that have been made for vertebrates. Such comparisons that can be made for the invertebrates are between phyla. The ancient geological age of the invertebrates has led to considerable diversity in their metabolism and this serves to obscure basic relationships. An additional complicating factor is that invertebrate phylogeny itself is not generally agreed upon.

Nevertheless, some tentative conclusions should be drawn at this point if for no other reason than to form a rationale for future studies. The general phylogenetic relationships of the terrestrial invertebrates discussed here are shown in Fig. 1. Because

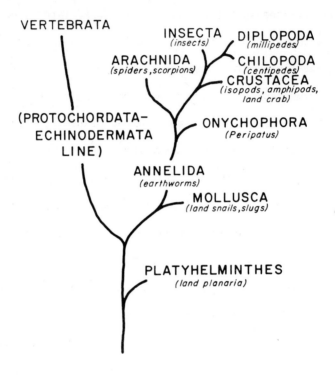

Fig. 1. Phylogeny of the major groups of terrestrial invertebrates.

the ornithine-urea cycle is also the source of arginine for nutritional uses, it is likely that the arginine-synthesizing part of the cycle developed early during biochemical evolution (21, 23). Land planaria are the most primitive terrestrial animal studied to date and are therefore as close an approximation to the first metazoan species to invade the land environment as is available.

It is therefore significant that at least one species, *Bipalium kewense*, has developed a ureotelic type of nitrogen metabolism. It is not possible, at this time, to even speculate as to the factors operating to bring about the transition from the ammonotelism of the fresh-water of marine ancestors of the land planarian to the ureotelism of extant species because of insufficient knowledge of the physiology of land planarians, especially their water relationships. The potential for urea synthesis must have persisted through the echinoderm-protochordate line of evolution that ultimately leads to the vertebrates since this potential is presumed to have been present in piscine ancestors of the lower vertebrates (19). The capacity for urea synthesis also persisted into the line of evolution leading to the annelids, the line that ultimately gave rise to the arthropods. In the molluscs, a group that constitutes a branch from the main annelid-arthropod line, urea synthesis takes place in land snails even though these forms are predominantly purinotelic. It seems unlikely that urea synthesis persisted through to the arthropods. The onychophoran *Peripatus* shares certain morphological features with both annelids and arthropods and has been considered by some to be the "missing link" between these two phyla. This organism appears to be uricotelic in its excretory nitrogen metabolism (84) and, in this respect, resembles more the arthropods than the annelids. The crustacean arthropods also appear incapable of urea synthesis. The ornithine-urea cycle is absent from the terrestrial isopod *Oniscus* (57) and the crayfish *Orconectes*. Although there is an increase in urea in the latter species when it is placed in a hypertonic environment, the metabolic origin of the urea appears not to be via the ornithine-urea cycle (111). Insects appear to have lost the capacity for urea synthesis in much the same way as birds. Arginine is a nutritional requirement for all insects studied (36) and we have been unable to demonstrate two enzymes of the ornithine-urea cycle, ornithine transcarbamylase and argininosuccinate lyase, in three insects using techniques that have been used to detect the same two enzymes in other invertebrates (99, 105). However, the pathway from citrulline to arginine and urea may be present in other insect species. This conversion has been shown in the silkworm *Bombyx mori* (70) and serves to explain the replacement by citrulline of the dietary arginine requirement in some insects (67). Insects thus present an interesting parallel with birds since both appear to have retained the same portion of the ornithine-urea cycle during their evolutionary transition to uricotelism (119).

Because the presence of arginase activity in invertebrates has sometimes been interpreted as evidence for urea synthesis *de novo*, the alternate metabolic function of arginase deserves some consideration. Some arthropods, like many other invertebrates (48), possess arginase activity. Arginase is present in insects (106) and isopods (57) but may be absent from crustaceans (110). Arginase is also present in several uricotelic vertebrates incapable of arginine synthesis (17, 19). In ureotelic vertebrates such as the mammal, arginase also has a catabolic role in the conversion of arginine to glutamate or proline in addition to its role in urea synthesis. The importance of this catabolic role is illustrated by the retention and control of arginase by mammalian liver cells in tissue culture even though these cells have lost a functional ornithine-urea cycle (42). The special catabolic function of arginase in proline formation described in insect flight muscle (105) may thus also prove to be significant in other invertebrates in which arginase occurs in the absence of a functional ornithine-urea cycle. During silkmoth development there is an increase in arginase activity coincident with the emergence of the winged adult (Fig. 2). Proline is utilized as a major substrate

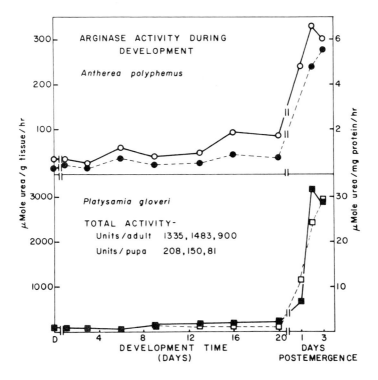

Fig. 2. Arginase activity during silkmoth development. Redrawn from (105).

in insect flight metabolism (109) because proline, but not glutamate, penetrates flight muscle mitochondria and undergoes oxidation via proline dehydrogenase to provide intermediates of the tricarboxylic acid cycle allowing for the complete oxidation of pyruvate (56, 108). The increase in arginase in the silkmoth coincident with flight can therefore be attributed to the function of the enzyme in converting arginine, formed by protein catabolism, to proline for flight as shown in Fig. 3 (105). The other

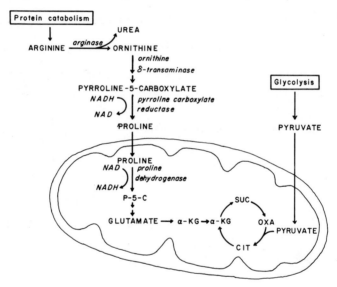

Fig. 3. Proposed role of arginase in insect flight metabolism.

enzymes involved in this conversion, ornithine δ-transaminase and Δ^1-pyrroline-5-carboxylate reductase, are also present in the silkmoth and both are extramitochondrial enzymes in keeping with their proposed role in proline formation for mitochondrial penetration. In all other animal tissues examined thus far, ornithine δ-transaminase is a mitochondrial enzyme (60, 118, 127). In mammalian liver, the major site of urea synthesis, ornithine δ-transaminase functions to form ornithine from glutamate for arginine synthesis; in kidney, which is highly gluconeogenic, the enzyme functions to form glutamate or proline from arginine (128).

Historically, terrestrial snails have occupied a central position in the comparative biochemistry of invertebrate nitrogen metabolism. Although primarily purinotelic, they were the first invertebrates in which key enzymes of the ornithine-urea cycle were demonstrated. The demonstration of arginase in the garden snail *Helix pomatia* (32) led to the first studies of the possible role of this enzyme in land snails (5). The demonstration of ornithine transcarbamylase and argininosuccinate lyase in a related snail *Otala lactea* (81) served to renew interest in the possible synthesis of arginine and urea by these invertebrates, both in our and other laboratories. The metabolism of land snails has been enigmatic in terms of classical comparative biochemistry. The demonstration of arginase in *Helix* was contrary to the "rule of Clementi" (31; *cf*. 17) that high levels of arginase occur only in the livers of ureotelic species. The presence of urease activity in certain land snails (97) has also been puzzling since this enzyme in invertebrates is classically associated with uricolysis (45) and yet uric acid is their major excretory product. Because some of the recently-described metabolic conversions in land snails may have broader implications in invertebrate nitrogen metabolism, we would like to discuss in detail certain aspects of the unique metabolism of urea in these snails, especially as it may relate to the physiological adaptation of these organisms to the terrestrial environment.

ENZYMES OF ARGININE BIOSYNTHESIS IN TERRESTRIAL SNAILS

Ornithine transcarbamylase and argininosuccinate lyase were first detected in an invertebrate animal in the land snail *Otala lactea* (81). The properties of these two molluscan enzymes appear to be similar to their vertebrate counterparts (27, 81). Ornithine transcarbamylase, argininosuccinate lyase and, in addition, argininosuccinate synthetase are now known to occur in several terrestrial and amphibious snails (Table 1) as well as in some aquatic species (47, 64). The levels of these enzymes in land snails are usually lower than those found in amphibians and mammals but, in some cases, are comparable to those in lower vertebrates such as the African lungfish (74). The activity of argininosuccinate synthetase is generally lower than that of ornithine transcarbamylase and argininosuccinate lyase. In vertebrates, the synthetase may be rate-limiting in arginine biosynthesis (20). There is some question as to whether the conditions employed for the assay of the snail enzyme give a true estimate of its activity. Although under the conditions used, there is a linear product formation with time and enzyme concentration (27), the levels measured in *Otala* and *Helix* hepatopancreas tissue will not account for the rate of conversion of citrulline sometimes observed in intact tissue (116). Extracts of the tissue are quite inhibitory for mammalian argininosuccinate synthetase (27) and this reaction

Table 1. *Enzymes of arginine biosynthesis in hepatopancreas tissue of terrestrial and amphibious snails*[a]

Species	Enzyme			
	CAP[b] synthetase	OTC	ASA synthetase	ASA lyase
PULMONATES (terrestrial):				
Otala lactea	+	+	+	+
Helix aspersa	+	+	+	+
H. pomatia	ND	+	ND	ND
Strophocheilus oblongus	+	+	+	+
Achatina fulica	ND	+	+	+
Rumina decollata	ND	+	+	+
Bulimulus dealbatus	ND	+	+	+
Limax maximus	ND	+	ND	ND
Euglandina singleyana	ND	+	ND	ND
Mesodon roemeri	ND	+	+	+
PROSOBRANCHS (aquatic-amphibious):				
Pila ampullaria	ND	+	+	+
Pomacea depressa	ND	+	+	+
Marsia cornaurietis	+	+	+	+
Sinotaia ingallsiana	+	+	+	+

[a] Compiled from (15, 27, 64, 94, 123, 124, 125).

[b] The abbreviations are: CAP, carbamyl phosphate; OTC, ornithine transcarbamylase; and ASA, argininosuccinic acid.

might thus be a point of regulation of the arginine pathway in the snails. The levels of the enzymes do not show much variation from one tissue to the other and there is thus little tissue differentiation with respect to the arginine pathway (27). This, plus the capacity of all tissues to synthesize arginine (26), may be indicative of a general tissue independence in invertebrates which is due, perhaps, to the lower efficiency of an open circulatory system as opposed to a closed one for intertissue communication.

Attempts to measure a carbamyl phosphate-synthesizing activity in terrestrial pulmonate snails were, until recently, negative or inconclusive (23). Carbamyl phosphate synthetase activity was reported in two prosobranch molluscs (64; Table 1) but the requirements for this activity were not established. The carbamyl phosphate synthetase in *Otala, Helix* and *Strophocheilus* is a unique enzyme (123, 124) whose properties distinguish it from either of the vertebrate enzymes, carbamyl phosphate synthetase-I or carbamyl phosphate synthetase-II (75). Vertebrate carbamyl phosphate synthetase-I is a mitochondrial enzyme that utilizes ammonia, carbon dioxide and ATP for carbamyl phosphate synthesis and requires N-acetyl-L-glutamate as a co-factor. Carbamyl phosphate synthetase-II is an extramitochondrial enzyme that utilizes L-glutamine, carbon dioxide and ATP but does not require a co-factor. As shown in Table 2, the carbamyl phosphate synthetase present in *Strophocheilus* hepatopancreas utilizes L-glutamine, carbon dioxide and ATP for carbamyl phosphate synthesis but, in addition, shows an absolute requirement for N-acetyl-L-glutamate. The requirement

Table 2. *Requirements for carbamyl phosphate synthesis by lysed mitochondrial preparations from* Strophocheilus oblongus *hepatopancreas*

Assay system modifications	Enzyme activity[a] (dpm citrulline/20 min)
None	20,051
Minus L-glutamine	108
Minus N-acetyl-L-glutamate	186
Minus ATP	14
Minus L-glutamine; plus 12.5 mM NH$_4$Cl	2,574
Minus L-glutamine and N-acetyl-L-glutamate; plus 12.5 mM NH$_4$Cl	2,136

[a]Measured as the incorporation of ^{14}C-bicarbonate into citrulline in the presence of L-ornithine and excess ornithine transcarbamylase. The L-glutamine and ^{14}C-bicarbonate concentrations used were 10 mM (123).

for this co-factor distinguishes the snail enzyme from vertebrate carbamyl phosphate synthetase-II (54, 55, 121). The failure to utilize low concentrations of ammonia distinguishes the snail enzyme from vertebrate carbamyl phosphate synthetase-I. The enzyme also differs from carbamyl phosphate synthetase-II in its behavior toward high concentrations of ammonia. Carbamyl phosphate synthetase-II will utilize ammonia in high concentration although it shows a much greater affinity for L-glutamine (54, 79, 85, 132). Some fixation of ^{14}C-bicarbonate into citrulline is obtained with the snail enzyme in the presence of ammonia (Table 2) but varying the NH_4^+-NH_3 ratio in the pH range 6.5 to 9 (Fig. 4) does not result in a typical pH optimum as it does for the vertebrate enzyme (54, 132). That glutamine is the substrate for the snail enzyme is evidenced in the

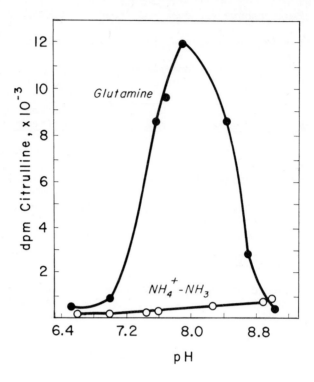

Fig. 4. Effect of pH on *Strophocheilus oblongus* carbamyl phosphate synthetase with L-glutamine of NH_4^+-NH_3 as substrates. From (123). (Reproduced by permission from the Editors of the Journal of Biological Chemistry.)

enzyme's inhibition by glutamine analogs. In concentration ratios with L-glutamine of 50 to 1, azaserine and O-carbamyl-L-serine inhibit the snail enzyme 20 and 80%, respectively. This inhibition is similar to that obtained with carbamyl phosphate synthetase-II from Ehrlich ascites tumor cells (54). The binding of L-glutamine by the snail enzyme is much less than that by the vertebrate enzyme. The K_m for L-glutamine of the snail carbamyl phosphate synthetase is around 2.5 mM as opposed to a K_m of around 0.01 mM for carbamyl phosphate synthetase-II (54, 79, 132). Glutamine is a major free amino acid in the snails, occurring in concentrations of from 1 to 2 µmoles per g (26). The high K_m of the snail carbamyl phosphate synthetase is thus within a physiologically functional range. L-Asparagine, a major dietary amino acid for the herbivorous snails, will not serve as a substrate for *Strophocheilus* carbamyl phosphate synthetase. Other properties of the snail enzyme that it shares in common with the two vertebrate enzymes are qualitative and not necessarily quantitative ones. There are, for example, differences between the invertebrate and vertebrate enzymes in their binding of ATP, Mg^{2+} and N-acetylglutamate (123).

The cellular localization of *Strophocheilus* carbamyl phosphate synthetase is shown in Table 3. The enzyme is highest in those fractions containing mitochondria (600 X g supernatant fluid and 6,600 X g residue). There is a low recovery of the activity following sedimentation of the mitochondria from the low-speed supernatant fluid and the subsequent solubilization and assay of the enzyme. This low recovery is presumably due to the instability of the enzyme although during lysis of the mitochondria by sonication, both glycerol and dithiothreitol are added as stabilizing agents (123). Following sedimentation in a sucrose density gradient, the carbamyl phosphate synthetase activity coincides with the activity of cytochrome *c* oxidase, a common mitochondrial marker enzyme (Fig. 5) lending additional support to the mitochondrial localization of the enzyme.

Table 3. *Cellular localization of carbamyl phosphate synthetase activity in* Strophocheilus oblongus *hepatopancreas*

Fraction	Enzyme activity	
	Total activity (dpm citrulline/20 min)[a]	% of Total
Homogenate[b]	748,125	100
600 X g:		
residue	28,036	3.8
supernatant fluid	867,468	115.9
6,600 X g:		
residue	437,404	58.5
supernatant fluid	24,624	3.3
110,000 X g:		
residue	64	0.01
supernatant fluid	37,664	5.0

[a] Measured as the N-acetyl-L-glutamate-dependent fixation of ^{14}C-bicarbonate into citrulline in the presence of L-glutamine (123).

[b] From 5.7 g tissue.

A carbamyl phosphate synthetase similar to the enzyme characterized from *Strophocheilus* hepatopancreas has also been detected in mitochondrial preparations from the snails *Otala* and *Helix*, the earthworm *Lumbicus terrestris* and the land planarian *Bipalium* (124). This type of carbamyl phosphate synthetase thus appears to have a wide distribution among invertebrate animals. The earthworm has been reported to have a carbamyl phosphate synthetase-I-like activity in the soluble fraction of gut tissue (11). We have also detected an ammonia and N-acetylglutamate mediated synthesis in the mitochondria from this tissue which amounts to 40% of the glutamine and N-acetylglutamate mediated activity (124). Whether the use of

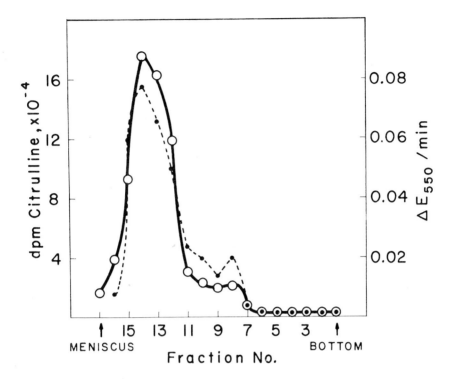

Fig. 5. Carbamyl phosphate synthetase and cytochrome *c* oxidase activities in *Strophocheilus oblongus* mitochondria sedimented in a sucrose density gradient. From (123). (Reproduced by permission from the Editors of the Journal of Biological Chemistry.)

different methods of preparation, especially the inclusion of glycerol and dithiothreitol as protective agents, results in modification of the enzyme's substrate binding properties such as occurs during purification of the glutamine enzyme from *Escherichia coli* (76) remains to be established.

NUTRITIONAL IMPLICATIONS

Since arginine is a universal constituent of protein, the absence of a functional arginine pathway from a species confers upon that species a dietary requirement for arginine or a precursor of this amino acid such a citrulline. This concept has been demonstrated both nutritionally and metabolically in the chicken (119). Considering the possible origin of the arginine pathway during biochemical evolution, it seems likely that the synthesis of protein arginine was of primary importance (21, 23). The utilization of the pathway for urea synthesis was a secondary development that allowed for the survival of metazoan animals, of which the land planarian is now a close relative of the most primitive forms, in water-poor environments. That the enzymes of the arginine pathway function for protein arginine synthesis has been demonstrated in the land snails *Otala*, *Helix* and *Strophocheilus* (26, 125). ^{14}C-Bicarbonate and [*ureido*-^{14}C]citrulline are incorporated into the *guanidino*-C of protein arginine by both whole animals *in vivo* or intact tissues *in vitro*. ^{14}C-Ornithine, variously labelled, gives rise to the ornithine carbons of protein arginine. A comparison of the specific radioactivities of the arginine, aspartate and glutamate in whole body protein following the injection of ^{14}C-bicarbonate shows them to be equivalent when allowance is made for the differing contents of each of these amino acids of the protein (Table 4). Because both aspartate and glutamate are seldom, if ever, required dietarily by animals and are formed by transamination reactions in the snails (25), these results indicate that the arginine pathway in the snails satisfies, in whole or part, their basic nutritional needs for L-arginine (26).

Table 4. *Incorporation of* ^{14}C-*bicarbonate into three protein amino acids by* Otala lactea *and* Helix aspersa

Amino acid	Protein content (mole %)	Specific radioactivity[a] (dpm/μmole)
Helix:		
Arginine	5.3	1971
Aspartate	10.3	1193
Glutamate	12.5	779
Otala:		
Arginine	5.3	3064
Aspartate	10.6	2129
Glutamate	12.9	1364

[a]Twelve hr after the injection of 20 μCi ^{14}C-bicarbonate (3 μmole) into individual *Helix* and 40 μCi (6 μmole) into individual *Otala*. From (26).

Just as the absence of a functioning arginine pathway has a basic nutritional implication, so does the absence of a functioning pyrimidine pathway. Since the carbamylation of aspartate to form carbamyl aspartate is the first step in pyrimidine biosynthesis (82), the absence of a carbamyl phosphate synthetase would confer upon a species a dietary requirement for pyrimidine in addition to the requirement for arginine. Dietary requirements for pyrimidine, if they occur at all, are uncommon among vertebrates and also insects (1), the latter being the only invertebrates whose nutrition has been studied in detail. The discovery of carbamyl phosphate synthetase-II in vertebrates, especially birds, was based on these nutritional considerations (14). The fixation of ^{14}C-bicarbonate into pyrimidines by pigeon liver homogenates (14), Ehrlich ascites tumor cells (53) and mouse spleen (72) subsequently led to the detection of the glutamine mediated synthesis of carbamyl phosphate in these tissues (54, 121). The discovery of a glutamine carbamyl phosphate synthetase in land snails has had a similar development. That carbamyl phosphate synthesis takes place in the tissues has been shown by the incorporation of ^{14}C-bicarbonate into the *ureido*-C of citrulline (24) and into the nucleic acid pyrimidines, UMP and CMP (96).

In ureotelic vertebrates, the two carbamyl phosphate synthetases, Enzyme-I and Enzyme-II, are thought to have separate metabolic functions (75). Enzyme-I is localized along with ornithine transcarbamylase in the mitochondrion and thus provides carbamyl phosphate for arginine and urea synthesis. Its high affinity for ammonia is ideally suited for the detoxication of ammonia formed through transdeamination via glutamate dehydrogenase. Enzyme-II occurs extramitochondrially in the cytosol along with aspartate transcarbamylase and provides carbamyl phosphate for pyrimidine biosynthesis. Although carbamyl phosphate formed intramitochondrially may be accessible to extramitochondrial enzyme systems (91), the cellular localization of the two carbamyl phosphate synthetases and the occurrence of Enzyme-II, but not Enzyme-I, in tissues capable of pyrimidine but not arginine synthesis *de novo* (54, 55, 121, 132) are convincing evidence for the two physiological functions of the enzymes.

The cellular localization of ornithine transcarbamylase and aspartate transcarbamylase in *Strophocheilus* hepatopancreas is shown in Table 5. The glutamine carbamyl phosphate synthetase in this

Table 5. *Cellular localization of transcarbamylase activity in* Strophocheilus oblongus *hepatopancreas*

Fraction	Enzyme activity			
	OTC[a]		ATC	
	Total units[b]	% of total	Total units	% of total
Homogenate[c]	1080	100	1.92	100
Nuclei, cell debris, etc. (600 X g residue)	398	36.8	0	0
Mitochondria (6,600 X g residue)	622	57.6	0.34	17.7
Soluble (110,000 X g supernatant fluid)	53	4.9	1.34	69.8

[a] The abbreviations are: OTC, ornithine transcarbamylase and ATC, aspartate transcarbamylase.

[b] A unit of activity corresponds to 1 μmole product/hr at 30°.

[c] From 4.5 g tissue. From (123).

tissue is localized mainly in the mitochondria (Table 3) along with ornithine transcarbamylase and not in the cytosol with aspartate transcarbamylase. This localization further distinguishes the snail carbamyl phosphate synthetase from vertebrate carbamyl phosphate synthetase-II. Because of its localization with ornithine transcarbamylase, the carbamyl phosphate synthetase in *Strophocheilus* hepatopancreas presumably functions for arginine biosynthesis. No definite carbamyl phosphate synthetase activity has been detected in the cytosol of this tissue although there is some non-specific fixation of ^{14}C-bicarbonate into citrulline in this fraction (123). There is thus the question of the origin of carbamyl phosphate for pyrimidine synthesis. The mitochondrial enzyme could have a dual function since intramitochondrially-formed carbamyl phosphate is accessible to extramitochondrial enzymes (91) and the glutamine enzymes in at least two microorganisms are known to function in both arginine and pyrimidine biosynthesis (see 71). Should this prove true, it would be of interest because of the proposed symbiotic origin of mitochondria (103): some molluscan species are now known to harbor free functional chloroplasts acquired by symbioses (126).

In addition to terrestrial snails, the function of the arginine pathway in providing protein arginine has been established in the earthworm (12) and the land planarian (21). Except for the snail *Helix* (96), the biosynthesis of pyrimidines in invertebrates has received little attention. Aspartate transcarbamylase occurs in the earthworm (12) and the biosynthesis of pyrimidines presumably takes place in this species as it does in vertebrates. The general absence of a dietary requirement for pyrimidines from insects and their similarities to birds with respect to the lack of a functional arginine pathway make them interesting species to examine for a pyrimidine-specific carbamyl phosphate synthetase.

ARGINASE AND UREASE

Arginase has an almost ubiquitous distribution among molluscs, occurring in at least some members of most classes (48). Terrestrial snails show the highest levels of enzyme activity. The enzyme occurs in the soluble cellular fraction of *Otala* hepatopancreas

(48). Arginase, like the enzymes of arginine biosynthesis, is present in several "extrahepatic" tissues (48). Several properties of the *Otala* enzyme are similar to those of mammalian liver arginase. The snail enzyme, for example, undergoes metal ion activation, shows a low K_m for L-arginine (2.5 mM) and is inhibited by L-ornithine (22). It differs from the mammalian enzyme in some of its molecular properties. For example, it may be distinguished from the mammalian enzyme by its migration during electrophoresis (22). The molecular weight of the *Otala* and *Helix* enzyme is 240,000 indicating that it is an octamer-type as opposed to the tetramer-type arginase that occurs in mammalian liver (107). In addition to arginase, guanidinobutyrate ureohydrolase also occurs in *Otala* and *Helix* (26, 95) but the metabolic function of this latter enzyme has not been investigated.

The enzyme urease was first reported in land snails by Przylecki (97). This observation was confirmed by Baldwin and Needham (5) for kidney but not hepatopancreas tissue. Baldwin and Needham failed to detect urease in the latter tissue because the enzyme is most active under alkaline conditions (59) and essentially inactive at the pH of their assay. Because most urea formed by hepatopancreas tissue is broken down by its urease, ornithine does not cause an increased accumulation of urea as it does with mammalian liver tissue and Baldwin and Needham were led to conclude that no urea synthesis took place in the land snail. The urease in land snails has now been recognized to play a central role in their metabolism of urea (23). Urease activity has been detected in hepatopancreas tissue of *Otala*, *Helix* and *Strophocheilus* (114, 125) and four additional terrestrial pulmonates (64). Whereas the enzyme has a very wide tissue distribution in *Otala* (114), it is restricted mainly to the hepatopancreas of *Strophocheilus*. The level of activity in *Strophocheilus* hepatopancreas is only about one-fourth that found in *Otala* hepatopancreas (125).

Although urease activity has been reported in several invertebrate species (see 112), there is considerable skepticism associated with reports of animal urease. The fact that the urease activity of higher vertebrates is of bacterial origin (77) has been confirmed in several fetal and germ-free animals (39). The enzyme from *Otala* hepatopancreas has now been purified 160-fold from acetone powders (87, 88) and there is no need to belabor the issue of its bacterial origin. There are, however, two points that can be made in this connection. First of all, the activity in *Otala* hepatopancreas is very constant from one individual to another. For example, the average value found for fourteen individuals was 39.7 ± 10.8 ($\pm$ SD) (88). One characteristic of mammalian gastric urease is its extreme quantitative and qualitative variability among individuals (39). Secondly, as shown in Table 6, the urease activity of *Otala* hepatopancreas is localized almost exclusively in the soluble cellular fraction. This is also true for *Strophocheilus* hepatopancreas urease (125). In tissues containing known ureolytic symbionts, sedimentation at 10,000 X g for 10 min removes 100% of the urease activity from the supernatant fraction (106).

Table 6. *Cellular localization of urease in Otala lactea hepatopancreas*

Fraction	Enzyme activity	
	Total units[a]	% of total
Homogenate[b]	39.0	100
Nuclei, cell debris, etc. (600 X g residue)	1.0	2.7
Mitochondria (6,600 X g residue)	0.3	0.7
Soluble (110,000 X g supernatant fluid)	41.4	106.7

[a] A unit of activity corresponds to 1 μmole urea hydrolyzed/hr at 30°.

[b] From 1 g tissue. From (88).

The K_m for urea of the enzyme purified from *Otala* hepatopancreas is 0.1 mM (88). This is ten times lower than the lowest reported K_m for urease which is that for "rumen contents" (100), thirty times lower than the K_m of plant urease (13) and three orders of magnitude lower than the K_m for urea of either invertebrate (112) or bacterial (2, 80) ureases. The lower K_m of the snail enzyme accounts for the rapid turnover of urea in the snails (see Fig. 7) despite their relatively low tissue urea contents. At a urea concentration of 0.2 mM, which is normal for *Otala* hepatopancreas, the urease in this tissue could hydrolyze ~27 μmoles urea/hr. Another distinctive property of the snail urease is its high pH optimum. The pH optimum for the *Otala* enzyme is in the range 8.5 to 9 depending upon the buffer used (59, 88, 114). The *Otala* enzyme differs from both plant and bacterial urease in its insensitivity to sulfhydryl reagents (88). For example, under conditions where the plant or bacterial enzymes are 100% inhibited by *p*-mercuribenzoate, the snail enzyme retains 98% of its activity. Similar results are obtained with other sulfhydryl reagents. The sensitivity of the snail urease to heavy metals is also quite different from that of the plant enzyme and does not follow the insolubility of their metallic sulfides. Silver or mercuric ions in 0.001 mM concentration are without effect on the snail enzyme although such concentrations inhibit the plant enzyme 100%. The snail enzyme does, however, show some similarities to the plant enzyme. It, for example, hydrolyzes hydroxyurea (44) and is inhibited by hydroxamates (43). The molecular weight of *Otala* urease is 262,000. This molecular weight corresponds to the "subunit" molecular weight of plant urease and to the molecular weight of *Bacillus pasteurii* urease (120).

INTEGRATION OF THE SYNTHETIC AND DEGRADATIVE PATHWAYS

In possessing both the biosynthetic and degradative pathways for arginine metabolism, land snails, at least *Otala* and *Helix*, are more like microorganisms (30, 101) than other animals. In bacteria that possess both systems, there may be a temporal compartmentation of the two systems. During growth, arginine synthesis occurs. When growth ceases, arginase is induced and arginine is then converted to glutamate (101). In *Neurospora crassa*, where the two systems also occur, it has been suggested that they are compartmented in such a way that the degradative system acts only on exogenously supplied arginine and not on biosynthetic arginine which is incorporated into protein (30). This compartmentation is attributed to the unique properties of the arginase present in *Neurospora* (90), the most important of which is the high K_m for arginine of the enzyme (0.1 to 0.2 M). Because the two systems in *Otala* are not temporally compartmented and the K_m for arginine of *Otala* arginase is a physiological one relative to the arginine concentration in most tissues (0.5 to 0.6 mM; ref. 26), unless there were a physical compartmentation separating arginase from the synthetic pathway it seemed possible that biosynthetic arginine would be converted to urea in spite of the subsequent hydrolysis of urea by urease. The integration of the two pathways in *Otala* hepatopancreas has been studied in two ways (116). In the first, the effect of acetohydroxamic acid, an inhibitor of *Otala* urease both *in vivo* and *in vitro* (88, 116), on the formation of $^{14}CO_2$ from precursors of ^{14}C-urea was determined. If reactions 1) through 4) were integrated in the two tissues, this inhibitor should affect the amount of $^{14}CO_2$ found.

1) [*ureido*-^{14}C]citrulline → [*guanidino*-^{14}C]argininosuccinate
2) [*guanidino*-^{14}C]argininosuccinate → [*guanidino*-^{14}C]arginine
3) [*guanidino*-^{14}C]arginine → ^{14}C-urea
4) ^{14}C-urea → $^{14}CO_2$

As shown in Table 7, acetohydroxamate inhibits almost completely the formation of $^{14}CO_2$ from ^{14}C-urea and its precursors. The second approach was to study the isotope dilution effect of ^{12}C-compounds

Table 7. *Effect of the urease inhibitor acetohydroxamic acid on the formation of $^{14}CO_2$ from labelled urea precursors by* Otala lactea *hepatopancreas*

	nMole $^{14}CO_2$/0.1 g tissue/hr		
Precursor	Tissue alone	Plus 5 mM acetohydroxamate	% Inhibition
[*ureido*-^{14}C]L-Citrulline	1.3	0.1	92.3
[*guanidino*-^{14}C]-Arginino-succinate	98.9	1.0	99.0
[*guanidino*-^{14}C]-Arginine	480.0	1.8	99.6
[^{14}C]Urea	390.0	2.8	99.3

From (116).

on $^{14}CO_2$ formation. If, for example, arginine were the intermediate in reactions 2) through 4), then the addition of ^{12}C-arginine in the presence of ^{14}C-argininosuccinate should decrease the amount of $^{14}CO_2$ formed. As shown in Fig. 6, this is the case. The net result of the integration of the biosynthetic and degradative pathways in *Otala* is the constant turnover of both arginine and urea in the snails (Fig. 7).

The synthesis and degradation of urea in *Otala* and *Helix* results in the formation of ammonia, some of which is volatilized as gaseous ammonia (114). ^{14}C-Carbon dioxide is volatilized from the snails following injection of ^{14}C-urea or its precursors (116) and, following the injection of ^{15}N-urea, the ammonia liberated is enriched with ^{15}N (114). Although "unknown symbiotic organisms" may contribute to the observed metabolism of urea in the snails, extensive

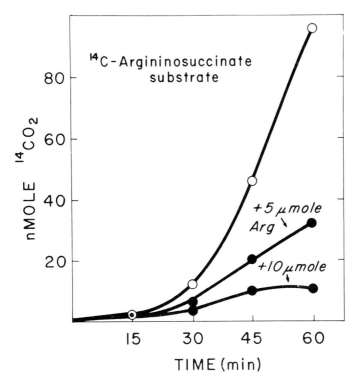

Fig. 6. Effect of ^{12}C-arginine on the formation of $^{14}CO_2$ from ^{14}C-argininosuccinate (guanidino labelled) by intact *Otala lactea* hepatopancreas tissue. Redrawn from (116).

treatment with a broad spectrum of antibiotics does not quantitatively affect its turnover. Assuming that purines are the major excretory products of *Otala*, we have attempted to assess the relative importance of gaseous ammonia production by the snails in their excretion of nitrogen (115). During estivation of the snails, our best estimate is that gaseous ammonia accounts for no more than one-third of the total nitrogen "excreted" if this total consists solely of ammonia-N plus purine-N.

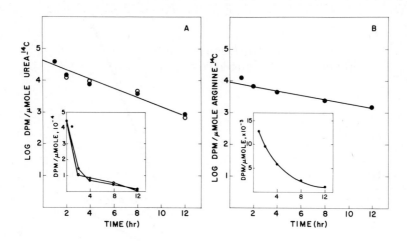

Fig. 7. Turnover of injected ^{14}C-urea and ^{14}C-arginine (guanidino labelled) in individual *Otala lactea*. The open circles refer to antibiotic-treated snails (116). From (23). (Reproduced by permission from Excerpta Medica Foundation.)

PHYSIOLOGICAL CONSIDERATIONS

From the foregoing considerations, it seems unlikely that urea formation is involved in ammonia detoxication in the snails. In vertebrates, glutamate dehydrogenase is a key enzyme in this process. The amino group of most amino acids is ultimately converted to glutamic acid by transdeamination. Glutamate penetrates the mitochondrion where the amino group is released as ammonia by glutamate dehydrogenase. The ammonia released may function in intramitochondrial pH regulation, required due to proton translocation (89). The excess ammonia is bound by carbamyl phosphate synthetase-I, an enzyme with an extremely high affinity for ammonia. Glutamate dehydrogenase thus has a central role in mammalian mitochondrial metabolism. We have, however, been unable to detect glutamate dehydrogenase in either the mitochondria or the cytosol of the three land snails *Otala*, *Helix* and *Strophocheilus* (41). As shown in Fig. 8, our

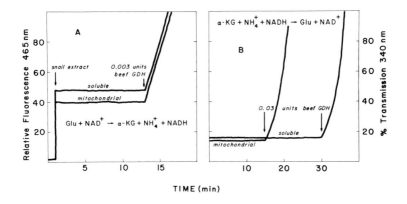

Fig. 8. Assay for glutamate dehydrogenase in *Otala lactea* hepatopancreas tissue. The tissue was homogenized in 9 vol. 0.25 M sucrose containing 50 mM phosphate buffer, pH 7.6. Fractionation by centrifuging was as described by Reddy and Campbell (105). The isolated mitochondria were suspended in the 50 mM phosphate buffer and lysed by sonication. Mitochondrial extract or soluble fraction equivalent to 10 mg original tissue wt. was used. NADH was measured spectrofluorometrically in the direction of NH_4^+ formation in the assay system of Strecker (117). Excitation was at 340 nm and the emission at 465 nm was monitored. A

assay conditions were adequate for detecting low levels of beef glutamate dehydrogenase or, not shown, the enzyme present in rat liver mitochondria. The presence of glutamate dehydrogenase in other molluscs is also questionable (25). The apparent absence of this enzyme is in contrast to the earthworm, a species in which ammonia detoxication may be served by urea synthesis and that has high levels of mitochondrial glutamate dehydrogenase (41). The general absence of glutamate dehydrogenase may explain the plethora of amino acid oxidases in molluscs (25). Rather than handling amino acid catabolism by transdeamination, molluscs may rely on catabolism via the oxidases. Transaminases are present in land snails and other molluscs (3, 25) but their cellular localization is not known. They may simply function to convert amino acids for which oxidases are lacking to those for which there are oxidases. The extremely low level of glutamine synthetase in mitochondrial extracts of *Strophocheilus* hepatopancreas (123) versus the high rate of glutamine formation observed *in vivo* by *Otala, Helix* and *Bulimulus alternata* (26) indicates that the major glutamine synthetase activity may be present in the cytosol although we have been unable to detect this activity with assays in which the activity is measured as glutamyl hydroxamate

---

(Fig. 8 [p. 33]) relative fluorescence of 100% equals 8.6 nmole NADH/ml. The assay system described in Bergmeyer (10) was used in the direction of glutamate formation. A unit of activity of beef glutamate dehydrogenase (GDH) corresponds to 1 μmole/min at 30°.

formation from glutamate and hydroxylamine (Speeg and Campbell, unpublished). Were this activity present in the cytosol, the ammonia formed by the action of amino acid oxidases could be converted to glutamine for mitochondrial penetration such as occurs in mammalian kidney (78). Thus the mitochondrial glutamine-dependent carbamyl phosphate synthetase in the land snails could be considered as a strictly synthetic enzyme for arginine and urea (and possibly pyrimidine) biosynthesis. Additional support for this idea comes from a consideration of the cellular localization of ornithine δ-transaminase in snail hepatopancreas tissue as opposed to its localization in insect fat body (Table 8). In the insect, where ornithine δ-transaminase plays a catabolic role in the conversion of arginine to proline (Fig. 3), the enzyme occurs solely in the cytosol. In snail hepatopancreas, on the other hand, the enzyme is mainly mitochondrial. Although mammalian kidney mitochondrial ornithine δ-transaminase may have a catabolic role (128), it is difficult to envisage such a role for the snail mitochondrial enzyme in the apparent absence of glutamate dehydrogenase (see Fig. 3). It is of interest in this connection that neither glutamate nor proline are oxidized by tissue preparations from several molluscs (25). Ornithine δ-transaminase in the snails may thus function in a strictly synthetic role in providing ornithine for arginine biosynthesis as it does in mammalian liver mitochondria (128). Ornithine is synthesized *in vivo* by *Otala*, presumably via glutamate (26).

The critical unresolved question concerning urea metabolism in terrestrial snails such as *Otala* and *Helix* is its turnover.

Table 8. *Cellular localization of ornithine δ-transaminase in three land snails and the insect* Hyalophora gloveri

Fraction[a]	% Total units			
	Otala lactea	Helix aspersa	Strophocheilus oblongus	Hyalophora[b] gloveri
Homogenate	100	100	100	100
Nuclei, cell debris, etc. (500 X g residue)	3.5	14.9	27.0	0
Mitochondria (15,000 X g residue)	73.1	64.4	51.2	0
Soluble (40,000 X g supernatant fluid)	9.6	17.7	0	103

[a]Tissue fractionation and the assay of ornithine δ-transaminase were as described by Reddy and Campbell (105). The total units (μmole/hr) per g tissue were as follows: for *Otala*, 26.8; for *Helix*, 42.3; for *Strophocheilus*, 67.5; and for *Hyalophora*, 66.7

[b]Data from (105).

What is the function of synthesizing urea with the expenditure of energy to only hydrolyze it back to ammonia and carbon dioxide? The answer to this question may ultimately reside in the role of the reaction $NH_3 + H^+ \rightleftarrows NH_4^+$ in both intra- and extracellular pH regulation. Intracellularly, proton translocation across the mitochondrial (or chloroplastal) membrane, which results in a transmembrane pH gradient, is central to the chemiosmotic hypothesis for phosphorylation (89; see also 98). Whereas "ammonia toxicity" may be defined at the cellular level in terms of the uncoupling effect of excess ammonia/ammonium ion on this process (16), some controlled release of ammonia may nevertheless be essential for intracellular pH regulation. Enzymes such as glutamate dehydrogenase or glutaminase (78) could play such a role in mammalian cells. As we have shown here, neither *Otala*, *Helix* nor *Strophocheilus* appear to possess a glutamate dehydrogenase. Glutaminase activity is also not detectable in either lung or hepatopancreas tissue of *Otala* (113). In the absence of either of these enzymes, it may thus be metabolically expedient for the snails to synthesize urea for the controlled release of ammonia via the urease reaction. Any "detoxication of ammonia" that occurs in the snails must take place by the glutamine synthetase reaction since their carbamyl phosphate synthetase does not appear to use ammonia as a substrate and glutamine synthetase is the first enzyme for ammonia detoxication via purine biosynthesis. The snails are herbivorous and their requirements for detoxifying excess ammonia from protein catabolism are minimal. That purine biosynthesis could account for their handling of excretory nitrogen can be illustrated by the following considerations. On a cabbage diet, *Otala* takes in an average of about 1 g per day corresponding to 150 μatoms protein N (48). The rates of purine biosynthesis observed *in vivo* for *Otala* would provide for the conversion of from 200 to 400 μatoms N per day to excretory purines by an average size (10 g) snail (115).

Because the snails deposit and maintain a calcium carbonate shell, extracellular pH regulation is especially critical to them. As has been summarized by Hodges and Lörcher (63) for the avian eggshell-forming system, a fundamental problem in calcium carbonate deposition is the mechanism for the neutralization of the proton formed by the dissociation of bicarbonate. Our model for the involvement of ammonia in the extracellular deposition of calcium carbonate is shown in Fig. 9 (28, 116).

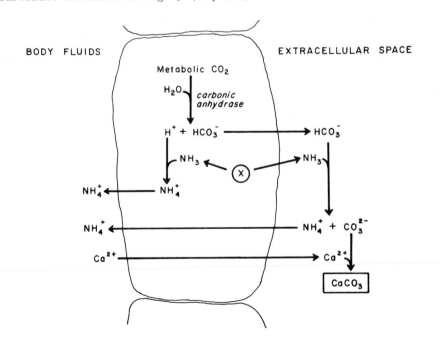

One of the perplexing problems in attempting to formulate a generalized scheme of metabolism for land snails has been that of their apparent species differences. For example, in comparing the metabolism of intact hepatopancreas tissue from *Otala* with that of the related snail *Bulimulus alternata*, we found that the latter had a rather straightforward "ureotelic-type" of metabolism in that the ^{14}C-urea formed from ^{14}C-arginine accumulated rather than being broken down by urease (26). Urease activity has, however, been detected in hepatopancreas tissue of the closely related snail *Bulimulus dealbatus* (64). ^{14}C-Urea synthesis from ^{14}C-bicarbonate has also been found in this latter species and, during estivation, there is an *accumulation* of urea (65). A similar synthesis and accumulation of urea occurs in two other terrestrial snails, *Mesodon roemeri* and *Rumina decollata*, and also the fresh-water snail *Helisona trivolvis* (66). Urea accumulation by certain tissues (hepatopancreas, lung and kidney) of *Strophocheilus* and *Thanusartus achilles* during estivation has also been reported (37). Since we have found urease activity in the hepatopancreas of *Strophocheilus* (10 μmole/g tissue/hr at 30°) and also a rapid turnover of injected ^{14}C-urea in actively feeding snails (125), we can only conclude that whatever the metabolic role of urea turnover, it appears to be regulated during different physiological states. The tissue distribution of urease

---

Fig. 9 [p. 38]. Proposed role of ammonia in calcium carbonate deposition. Modified from (28). The penetration of NH_4^+ through membranes is presumed to be as $NH_3 + H^+$.

in *Strophocheilus* appears to be quite different from that in *Otala*: whereas in the former species, urease is mainly restricted to the hepatopancreas with only trace amounts in the kidney, in *Otala*, urease has a ubiquitous tissue distribution (114). Unlike the *Otala* enzyme, the hepatopancreas urease of *Strophocheilus* is sensitive to EDTA although we have not found a cation requirement for the enzyme (Tramell, unpublished). Neither *Otala* nor *Helix* have ever been observed to accumulate urea under any condition (66, 113).

The accumulation of urea for the prevention of evaporative water loss by terrestrial and amphibious snails would be highly significant in terms of their adaptation to the terrestrial environment. That terrestrial snails undergo profound changes in water content has been known for some time (68, 129). The evaporation of water from *Helix aspersa* is almost identical to that from a free water surface (83). The accumulation of urea would thus decrease evaporative water loss by decreasing the vapor pressure of the body fluids. Such a mechanism might allow for survival of amphibious snails, such as the prosobranch vectors of human schistosomiasis *Pomatiopsis* and *Oncomelania*, during drought conditions. Van der Schalie and Getz (quoted in ref. 69) consider this group of snails to be evolving into a completely terrestrial mode of existence. We have attempted to duplicate natural estivation in the laboratory by removing individual *Strophocheilus* from food and water. The weight loss during this deprivation is shown in Fig. 10. In two experiments, the total weight loss at the end of 52 and 96 days was about 25% of the original weight. When this weight loss is

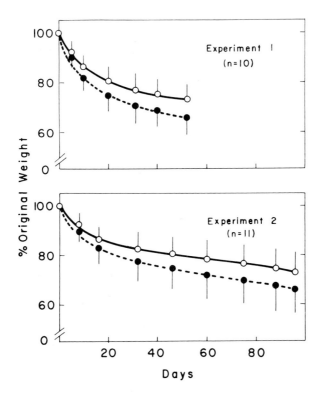

Fig. 10. Weight loss by *Strophocheilus oblongus* removed from food and water. The solid line represents the total body wt. The dashed line has been corrected for the shell wt. which remains constant (22.03 ± 2.6 [SD,n=4]% of the total wt. of snails at 0 time). The vertical lines represent the magnitude of the SD.

corrected for the shell weight, which remains constant, the decreased weight of the soft parts was about 35% that of the original weight. This weight loss appears to be due mainly to water loss from the blood and not to tissue dehydration. This is based on the data in Table 9. The percentage dry weight of four tissues increases only

Table 9. *Changes in tissue dry weight and blood protein contents in* Strophocheilus oblongus *during food and water deprivation*

Tissue	% Dry wt.[a]	
	Active snails	Snails deprived 52 days
Hepatopancreas	27.2 ± 3.6 (5)[b]	31.3 ± 5.0 (5)
Albumen gland	38.0 ± 3.5 (4)	41.0 ± 1.8 (3)
Reproductive tract	24.3 ± 0.5 (4)	24.5 ± 2.3 (5)
Foot	15.5 ± 2.9 (5)	16.9 ± 2.9 (5)
	mg Protein/ml[c]	
Blood	21.5 ± 7.7 (5)	46.8 ± 20.4 (5)

[a] After 24 hr at 115°.
[b] Values ± SD. Number of individuals in parenthesis.
[c] Lowry protein (bovine serum albumen standard).

1 to 2%, if at all, whereas the blood protein content almost doubles. As shown in Fig. 11, during deprivation, there is a marked accumulation of urea in the tissues of some snails but not others. The blood urea contents show the most consistent increase in urea. In one case, after 96 days deprivation, the blood urea content was 2.64% which is in the range of elasmobranch blood. DeJorge and Petersen (37) report higher hepatopancreas urea values for "hibernating" and "estivating" *Strophocheilus*. In hibernating snails, they report a high mean value of 311 μmoles/g after 90 days; 370 μmoles/g is reported for estivating snails after only 30 days. In

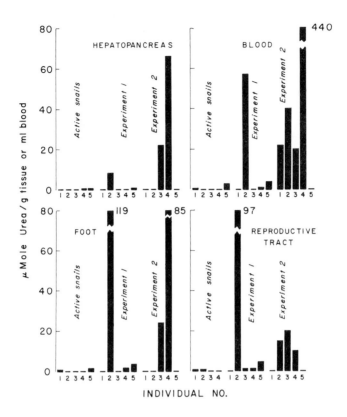

Fig. 11. Urea contents of active and deprived *Strophocheilus oblongus*. Five actively feeding individuals and five individuals from each experiment described in Fig. 10 were analyzed for urea. The tissues were homogenized in 9 vol. 0.5 M perchloric acid and the protein precipitate removed by centrifuging. The supernatant fluid was neutralized with potassium hydroxide and divided into two portions after the potassium perchlorate was removed by centrifuging. One portion was treated with urease to serve as a blank correction. Urea was then determined in each portion colorimetrically (81).

spite of the accumulation of urea in the blood and tissues of *Strophocheilus*, some urea turnover must still be taking place in the tissues. This is based on the apparent failure of the blood urea to equilibrate with the tissues. For example, snail #4 in *Experiment 2* shows a blood urea content of 440 μmole/ml. The hepatopancreas urea content of this snail is only about 100 μmole/ml tissue water, based on the tissue dry weights (Table 9). The reproductive tract tissue from snail #4 shows a blood urea/tissue urea ratio of 14. These results raise a question concerning the observed tissue distribution of urease in this species based on enzyme assays *in vitro*.

The mechanism of urea accumulation in terrestrial and amphibious land snails, especially as it relates to the occurrence, origin and possible regulation of urease, remains to be established. In *Otala* and *Helix*, where there is convincing evidence that the urease is not of bacterial origin, urea accumulation has not been observed under laboratory conditions. If urease is a constitutive enzyme of those species that accumulate urea, its regulation to allow such accumulation during periods of desiccation would be a very significant mechanism in their physiological adaptation to the terrestrial mode of existence, equivalent in importance to the accumulation of urea by elasmobranchs and lungfish in water-poor environments.

SUMMARY

Terrestrial and semi-terrestrial invertebrates are mainly purinotelic in their excretory nitrogen metabolism. Some also utilize a modified "ammonotelism" in which gaseous ammonia is volatilized directly into the environment. Ureotelism, although rare, does occur in earthworms under certain circumstances and in at least one land planarian indicating that this mode of nitrogen excretion has also been utilized by invertebrates in their evolutionary transition to the terrestrial environment.

Terrestrial snails, which are primarily purinotelic, are unique in possessing both the capacity to synthesize urea via the ornithine-urea cycle and an animal urease. The carbamyl phosphate synthetase present in these snails is a mitochondrial enzyme that utilizes L-glutamine, but not ammonia, as a substrate and has an absolute requirement for N-acetyl-L-glutamate. This enzyme is thus distinct from either vertebrate carbamyl phosphate synthetase-I or -II. The snails appear to lack glutamate dehydrogenase and their synthesis of urea may not be directed toward ammonia detoxication as it is in ureotelic vertebrates. Because of their tissue urease, there is a constant turnover of urea *in vivo* due to its synthesis and degradation. The urease present in the snails is distinct in its properties from either the plant or bacterial enzymes. The physiological role for urea turnover in the snails is postulated to be for both intra- and extracellular pH regulation. The latter is important in the process of depositing and maintaining a calcium carbonate shell. The turnover of urea appears to be regulated such that during water deprivation, urea can accumulate in the blood and to a lesser extent in the tissues. This would be of adaptive significance in preventing evaporative water loss. The synthesis and degradation of urea and the regulation of these processes are new aspects of the biology of terrestrial and amphibious snails that may prove to be of primary importance in their adaptation to the terrestrial environment.

REFERENCES

1. Altman, P.L. and D.S. Dittmer (editors). *Metabolism*. Bethesda, Fed. Amer. Soc. Exp. Biol., 1968. p. 162.
2. Andersen, J.A., F. Kopka, A.J. Siedler and E.G. Nohle. *Fed. Proc. 28*:764, 1969.
3. Awapara, J. and J.W. Campbell. *Comp. Biochem. Physiol. 11*:231, 1964.
4. Baldwin, E. *Comp. Biochem. Physiol. 1*:24, 1960.
5. Baldwin, E. and J. Needham. *Biochem. J. 28*:1372, 1934.
6. Balinsky, J.B. In: *Comparative Biochemistry of Nitrogen Metabolism. Vol. 2. The Vertebrates,* edited by J.W. Campbell. London and New York: Academic Press, 1970. p. 519.
7. Balinsky, J.B., M.M. Cragg and E. Baldwin. *Comp. Biochem. Physiol. 3*:236, 1961.
8. Balinsky, J.B., E.L. Choritz, C.G.L. Coe and G.S. Van der Schans. *Comp. Biochem. Physiol. 22*:59, 1967.
9. Baze, W.B. and F.R. Horne. *Comp. Biochem. Physiol. 34*:91, 1970.
10. Bergmeyer, U.H. (editor). *Methods of Enzymatic Analysis*. Weinheim: Verlag Chemie, 1963. p. 752.
11. Bishop, S.H. and J.W. Campbell. *Science 142*:1583, 1963.
12. Bishop, S.H. and J.W. Campbell. *Comp. Biochem. Physiol. 15*:51, 1965.

13. Blakeley, R.L., E.C. Webb and B. Zerner. *Biochemistry* 8:1948, 1969.
14. Bowers, M.D. and S. Grisolia. *Comp. Biochem. Physiol.* 5:1, 1960.
15. Bricteux-Grégoire, S. and M. Florkin. *Comp. Biochem. Physiol.* 12:55, 1964.
16. Brierley, G.P. and C.D. Stoner. *Biochemistry* 9:708, 1970.
17. Brown, G.W., Jr. *Arch. Biochem. Biophys.* 114:184, 1966.
18. Brown, G.W., Jr. and S.G. Brown. *Science* 155:570, 1967.
19. Brown, G.W., Jr. and P.P. Cohen. *Biochem. J.* 75:82, 1960.
20. Brown, G.W., Jr., W.R. Brown and P.P. Cohen. *J. Biol. Chem.* 234:1775, 1959.
21. Campbell, J.W. *Nature* 208:1229, 1965.
22. Campbell, J.W. *Comp. Biochem. Physiol.* 18:179, 1966.
23. Campbell, J.W. In: *Urea and the Kidney*, edited by B. Schmidt-Nielsen. Amsterdam: Excerpta Medica Fdn., 1970. p. 48.
24. Campbell, J.W. and S.H. Bishop. *Biochim. Biophys. Acta* 77:149, 1963.
25. Campbell, J.W. and S.H. Bishop. In: *Comparative Biochemistry of Nitrogen Metabolism. Vol. 1. The Invertebrates*. London and New York: Academic Press, 1970. p. 103.
26. Campbell, J.W. and K.V. Speeg, Jr. *Comp. Biochem. Physiol.* 25:3, 1968.
27. Campbell, J.W. and K.V. Speeg, Jr. *Z. vergl. Physiol.* 61:164, 1968.
28. Campbell, J.W. and K.V. Speeg, Jr. *Nature* 224:725, 1969.

29. Carlisky, N.J., A. Barrio and L.I. Sadnik. *Comp. Biochem. Physiol. 29*:1259, 1969.
30. Castañeda, M., J. Martuscelli and J. Mora. *Biochim. Biophys. Acta 141*:276, 1967.
31. Clementi, A. *Atti R. Accad. Lincei. Rendic. 23*:612, 1914.
32. Clementi, A. *Atti R. Accad. Lincei. Rendic. 27*:299, 1918.
33. Cohen, P.P. *Harvey Lect. Ser. 60*:119, 1966.
34. Cohen, P.P. *Science 168*:533, 1970.
35. Cohen, S.S. and W.B. Lewis. *J. Biol. Chem. 180*:79, 1949.
36. Corrigan, J.J. In: *Comparative Biochemistry of Nitrogen Metabolism. Vol. 1. The Invertebrates*, edited by J.W. Campbell. London and New York: Academic Press, 1970. p. 387.
37. DeJorge, F.B. and J.A. Petersen. *Comp. Biochem. Physiol. 35*:211, 1970.
38. Delaunay, H. *Biol. Rev. 6*:265, 1931.
39. Delluva, A.M., K. Markley and R.E. Davies. *Biochim. Biophys. Acta 151*:646, 1968.
40. Denton, D.A., M. Reich and F.J. Hird. *Science 139*:1225, 1963.
41. Drotman, R.B. and J.W. Campbell, in preparation.
42. Eliasson, E.E. and H.J. Strecker. *J. Biol. Chem. 241*:5757, 1966.
43. Fishbein, W.N. and P.P. Carbone. *J. Biol. Chem. 240*:2407, 1965.
44. Fishbein, W.N., T.S. Winter and J.D. Davidson. *J. Biol. Chem. 240*:2402, 1965.
45. Florkin, M. and G. Dûchateau. *Arch. Internat. Physiol. 53*:267, 1943.
46. Florkin, M. and G. Frappez. *Arch. Internat. Physiol. 50*:197, 1940

47. Friedl, F.E. and R.A. Bayne. *Comp. Biochem. Physiol.* 17:1167, 1966.
48. Gaston, S. and J.W. Campbell. *Comp. Biochem. Physiol.* 17:259, 1966.
49. Gifford, C.A. *Amer. Zool.* 8:521, 1968.
50. Goldstein, L. and R.P. Forster. In: *Comparative Biochemistry of Nitrogen Metabolism. Vol. 2. The Vertebrates*, edited by J.W. Campbell. London and New York: Academic Press, 1970. p. 495.
51. Goldstein, L., P.A. Janssens and R.P. Forster. *Science* 157:316, 1967.
52. Gordon, M.S., K. Schmidt-Nielsen and H.M. Kelley. *J. Exp. Biol.* 39:659, 1961.
53. Hager, S.E. and M.E. Jones. *J. Biol. Chem.* 240:4556, 1965.
54. Hager, S.E. and M.E. Jones. *J. Biol. Chem.* 242:5667, 1967.
55. Hager, S.E. and M.E. Jones. *J. Biol. Chem.* 242:5674, 1967.
56. Hansford, R.G. and B. Sacktor. *J. Biol. Chem.* 245:991, 1970.
57. Hartenstein, R. *Amer. Zool.* 8:507, 1968.
58. Hartenstein, R. In: *Comparative Biochemistry of Nitrogen Metabolism. Vol. 1. The Invertebrates*, edited by J.W. Campbell. London and New York: Academic Press, 1970. p. 299.
59. Heidermanns, C. and I. Kirchner-Kühn. *Z. vergl. Physiol.* 34:166, 1952.
60. Hill, D.L. and P. Chambers. *Biochim. Biophys. Acta* 148:435, 1969.
61. Hill, L. and W.H. Dawbin. *Comp. Biochem. Physiol.* 31:453, 1969.
62. Hill, L. and W.H. Dawbin. *Nature* 224:1325, 1969.

63. Hodges, R.D. and K. Lörcher. *Nature 216*:609, 1967.
64. Horne, F. and V. Boonkoon. *Comp. Biochem. Physiol. 32*:141, 1970.
65. Horne, F.R. *Comp. Biochem. Physiol. 38A*:565, 1971.
66. Horne, F.R. and G. Barnes. *Z. vergl. Physiol. 69*:452, 1970.
67. House, H.L. In: *The Physiology of Insects, Vol. 2.*, edited by M. Rockstein. New York: Academic Press, 1965. p. 769.
68. Howes, N.H. and G.P. Wells. *J. Exp. Biol. 11*:327, 1934.
69. Hyman, L.H. *The Invertebrates. Vol. VI. Mollusca I.* New York: McGraw-Hill, 1967. p. 375.
70. Inokuchi, T., Y. Horie and T. Ito. *Biochem. Biophys. Res. Commun 35*:783, 1969.
71. Issaly, I.M., A.S. Issaly and J.L. Reissig. *Biochim. Biophys. Acta 198*:482, 1970.
72. Ito, K. and M. Tatibana. *Biochem. Biophys. Res. Commun. 23*:672, 1966.
73. Janssens, P.A. *Comp. Biochem. Physiol. 11*:105, 1964.
74. Janssens, P.A. and P.P. Cohen. *Science 152*:358, 1966.
75. Jones, M.E. In: *Urea and the Kidney*, edited by B. Schmidt-Nielsen. Amsterdam: Excerpta Medica, 1970. p. 35.
76. Kalman, S.M., P.H. Duffield and T. Brzozowski. *J. Biol. Chem. 241*:1871, 1966.
77. Kornberg, H.L. and R.E. Davies. *Physiol. Rev. 35*:169, 1955.
78. Kovačević, Z., J.D. McGiven and J.B. Chappell. *Biochem. J. 118*:265, 1970.
79. Lan, S.J., H.J. Sallach and P.P. Cohen. *Biochemistry 8*:3673, 196

80. Larson, A.D. and R.E. Kallio. *J. Bacteriol. 68*:67, 1954.
81. Linton, S.N. and J.W. Campbell. *Arch. Biochem. Biophys. 97*: 360, 1962.
82. Lowenstein, J.M. and P.P. Cohen. *J. Biol. Chem. 220*:57, 1956.
83. Machin, J. *J. Exp. Biol. 41*:759, 1964.
84. Manton, S.M. *Phil. Trans. Roy. Soc. (London) B227*:411, 1937.
85. Maresch, C.G., T.H. Kwan and S.M. Kalman. *Canad. J. Biochem. 47*:61, 1969.
86. McClanahan, L., Jr. *Comp. Biochem. Physiol. 20*:73, 1967.
87. McDonald, J.A. and J.W. Campbell. *Fed. Proc. 29*:904, 1970.
88. McDonald, J.A. and J.W. Campbell, in preparation.
89. Mitchell, P. *Fed. Proc. 26*:1370, 1967.
90. Mora, J., R. Tarrab and L.F. Bojalil. *Biochim. Biophys. Acta 118*:206, 1966.
91. Natale, P.J. and G.C. Tremblay. *Biochem. Biophys. Res. Commun. 37*:512, 1969.
92. Needham, A.E. In: *Comparative Biochemistry of Nitrogen Metabolism. Vol. 1. The Invertebrates*, edited by J.W. Campbell. London and New York: Academic Press, 1970. p. 207.
93. Pickford, G.E. and F.B. Grant. *Science 155*:568, 1967.
94. Porembska, Z. and J. Heller. *Acta Biochim. Polon. 9*:385, 1962.
95. Porembska, Z., I. Gasiorowska and I. Mochnacka. *Acta Biochim. Polon. 15*:171, 1968.
96. Porembska, Z., B. Gorzkowski and M.M. Jezewska. *Acta Biochim. Polon. 13*:107, 1966.
97. Przylecki, St. J. *Arch. Internat. Physiol. 20*:103, 1922.

98. Racker, E. In: *Membranes of Mitochondria and Chloroplasts,* edited by E. Racker. New York: Van Nostrand Reinhold, 1970. p. 127.
99. Raghupathiramireddy, S. and J.W. Campbell. *Amer. Zool.* 7:195, 1967.
100. Rahman, S.A. and P. Decker. *Nature* 209:618, 1966.
101. Ramaley, R.F. and R.W. Bernlohr. *Arch. Biochem. Biophys.* 117:34, 1966.
102. Rao, K.P. and T. Gopalakrishnareddy. *Comp. Biochem. Physiol.* 7:175, 1962.
103. Raven, P.H. *Science* 169:641, 1970.
104. Read, L.J. *Nature* 215:1412, 1967.
105. Reddy, S.R.R. and J.W. Campbell. *Biochem. J.* 115:495, 1969.
106. Reddy, S.R.R. and J.W. Campbell. *Comp. Biochem. Physiol.* 28:515, 1969.
107. Reddy, S.R.R. and J.W. Campbell. *Comp. Biochem. Physiol.* 32:499, 1970.
108. Sacktor, B. and C.C. Childress. *Arch. Biochem. Biophys.* 120:583, 1967.
109. Sacktor, B. and E. Wormser-Shavit. *J. Biol. Chem.* 241:624, 1966.
110. Sharma, M.L. *Comp. Biochem. Physiol.* 24:55, 1968.
111. Sharma, M.L. *Comp. Biochem. Physiol.* 30:309, 1969.
112. Simmons, J.E., Jr. *Biol. Bull.* 121:535, 1961.
113. Speeg, K.V., Jr. Masters Thesis, Rice University, Houston, Texas. 1966.
114. Speeg, K.V., Jr. and J.W. Campbell. *Amer. J. Physiol.* 214:1392, 1968.

115. Speeg, K.V., Jr. and J.W. Campbell. *Comp. Biochem. Physiol.* *26*:579, 1968.

116. Speeg, K.V., Jr. and J.W. Campbell. *Amer. J. Physiol. 216*:1003, 1969.

117. Strecker, H.J. In: *Methods in Enzymology*, edited by S.P. Colowick and N.O. Kaplan. Vol. II. New York: Academic Press, 1955. p. 220.

118. Swick, R.W., S.L. Tollaksen, S.L. Nance and J.F. Thomson. *Arch. Biochem. Biophys. 136*:212, 1970.

119. Tamir, H. and S. Ratner. *Arch. Biochem. Biophys. 102*:259, 1963.

120. Tannis, R.J. and A.W. Naylor. *Biochem. J. 108*:771, 1968.

121. Tatibana, M. and K. Ito. *Biochem. Biophys. Res. Commun. 26*: 221, 1967.

122. Tercafs, R.R. and E. Schoffeniels. *Life Sci. 1*:19, 1962.

123. Tramell, P.R. and J.W. Campbell. *J. Biol. Chem. 245*:6634, 1970.

124. Tramell, P.R. and J.W. Campbell. *Comp. Biochem. Physiol.* in press, 1971.

125. Tramell, P.R. and J.W. Campbell, in preparation.

126. Trench, R.K. *Nature 222*:1071, 1969.

127. Vecchio, D. and S.M. Kalman. *Arch. Biochem. Biophys. 127*:376, 1968.

128. Volpe, P., R. Sawamura and H.J. Strecker. *J. Biol. Chem. 244*: 719, 1969.

129. Wells, G.P. *J. Exp. Biol. 20*:79, 1944.

130. Wieser, W. and G. Schweizer. *J. Exp. Biol. 52*:267, 1970.

131. Wieser, W., G. Schweizer and R. Hartenstein. *Oecologia 3*: 390, 1969.

132. Yip, M.C.M. and W.E. Knox. *J. Biol. Chem.* 245:2199, 1970.

ADAPTATION OF UREA METABOLISM

IN AQUATIC VERTEBRATES[1]

Leon Goldstein

Division of Biomedical Sciences
Brown University
Providence, Rhode Island   02912

Aquatic vertebrates (fish and aquatic amphibia and reptiles) hold a unique position in the animal kingdom. They occupy an evolutionary niche between the more primitive invertebrates and the more advanced terrestrial vertebrates. Fish as well as aquatic amphibia and reptiles have invaded and remained in every aqueous (salt, fresh and brackish water) habitat conceivable with the exception of a few that are incompatible with all but the most primitive forms of life. In addition several fish and aquatic amphibia can live out of water for long periods of time. The ability of these aquatic vertebrates to live in such a variety of environmental conditions requires the adaptation of a variety of life supporting systems to these conditions. Regarding nitrogen metabolism specifically, two major adaptations have taken place. The first of these is the accumulation of nitrogenous end-products to achieve osmotic equilibrium with an

---

[1]Supported by NSF Grant GB 8200.

otherwise hyperosmotic environment. The second is the use of ammonia-fixing biochemical reactions for the detoxification of this poisonous base when it cannot be readily eliminated from the body.

These adaptations are most apparent when one compares the nitrogen metabolism of aquatic organisms living under different degrees of water availability (2, 34). The biosynthetic and, for the most part, degradative pathways of nitrogen metabolism are similar in all animals. However, specific differences do exist in the nature and formation of the end-products of nitrogen metabolism in different animals. The major end-products of protein metabolism are ammonia, urea, and uric acid (2). Ammonia has many advantages as an end-product of nitrogen metabolism. In contrast to urea and uric acid, no expenditure of energy is required for the conversion of protein nitrogen to ammonia. In addition the small size of the molecule and the highly lipid soluble nature of the free base permits its easy elimination by diffusion without an accompanying obligatory loss of water across the gills and body surface of aquatic organisms. However, ammonia does have the disadvantage of being highly toxic and cannot be stored as such in the body. This property precludes its retention for osmotic purposes or during periods of water deprivation when it cannot be excreted. Thus, aquatic organisms endeavoring to avoid water loss to an otherwise hyperosmotic environment by retaining high concentrations of nitrogenous end-products in their body fluids must convert ammonia formed during metabolism to a less toxic end-product, *e.g.*, urea. Similarly, during periods of water deprivation aquatic organisms estivating in the mud convert ammonia to urea and store it as such until it can be excreted when the environment is rehydrated.

Most lower vertebrates living in an aquatic environment are ammonotelic; their major nitrogenous end-product is ammonia. These include the teleost fishes (7, 12) and aquatic amphibia such as *Necturus* (8) and *Xenopus* (3, 27). However, a large groups of fishes, the elasmobranchs are ureotelic (12, 33). These fish retain high concentrations of urea in their body fluids (to achieve osmotic equilibrium with the surrounding seawater) by actively reabsorbing the compound from the glomerular filtrate (22, 33). The mechanism of urea formation in elasmobranchs as well as other fishes had been in doubt since the early studies of Manderscheid (23) on urea biosynthesis in lower vertebrates. She reported that the ornithine-urea cycle was not present in the fishes she examined. In addition Baldwin (1) reported that although four of the five enzymes of the cycle were present in the livers of elasmobranchs, the enzyme incorporating ammonia carbon dioxide and ATP into carbamoylphosphate, carbamoyl phosphate synthetase (CPS) was not detectable in these fish. However, recent studies using improved assay conditions (6, 37) and radioisotopic techniques (16) have shown that low levels of CPS are present in the livers of elasmobranchs and that the ornithine-urea cycle is functional in these fish (28). An alternate route for the formation of urea in lower vertebrates is the purine pathway in which purines are enzymatically degraded to urea (12). Florkin and Dûchateau (10) found high levels of uricolytic enzymes in fish liver and we (14) have obtained evidence to indicate that this pathway probably accounts for the small amounts of urea formed in teleost fish. It was possible, therefore, that this pathway might have been the source of urea in elasmobranchs also. Thus, we (28) compared the rate of formation of urea in the elasmobranch *Squalus acanthias* via the purine pathway with that synthesized via the ornithine-urea cycle. These experiments were conducted both *in vivo* and with liver slices *in vitro*. The activity of the ornithine-urea cycle was assayed by determining the rate of incorporation of ^{14}C-bicarbonate into urea, that of the purine pathway by following the rate of incorporation of 2-^{14}C-uric acid and 3-^{14}C-serine into urea. The activity of the purine pathway *in vivo* was 1/10 and *in vitro* was 1/100 that of the ornithine-urea cycle (Fig. 1). Thus the ornithine-urea cycle appears to be the major pathway of urea formation in elasmobranchs.

Although most elasmobranchs are marine, some periodically enter and remain in fresh and brackish water for different periods of time (18, 31, 33, 36). Little is known about the biochemical mechanisms that allow elasmobranchs to adapt to a freshwater environment. The early studies of Smith (31) on the sawfish, *Pristis microdon*, and the later investigations of Thorson (36) on the bullshark, *Carcharinus leucas* showed that the adaptation of these suryhaline elasmobranchs to freshwater is accompanied by marked reductions in urea concentrations in the body fluids. We (18) investigated the mechanism of urea reduction during environmental dilution in the euryhaline lemon shark *Negaprion brevirostris*. In any animal the concentration of urea in the body fluids is the resultant of the

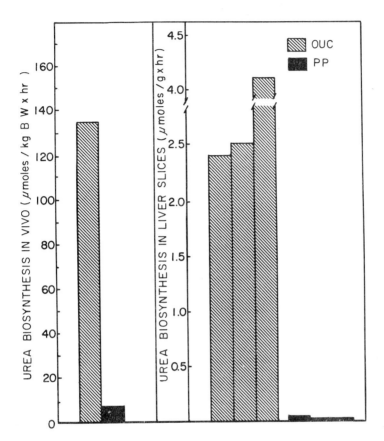

Fig. 1. Urea biosynthesis in the dogfish *Squalus acanthias*. OUC = ornithine-urea cycle. PP = purine pathway. Adapted from Schooler *et al.* (28).

difference between the rates of production and excretion of the compound. The rates of both parameters were measured by radioisotopic techniques in two groups of lemon sharks maintained either in 100% or 50% seawater. Although the rates of urea biosynthesis were similar in both groups, the rates of urea excretion were elevated threefold in the fish maintained in the dilute environment. In these fish, therefore, reduction in blood urea concentration following environmental dilution was due to an increased rate of excretion of the end-product. The mechanism(s) of the increased excretion of urea was not investigated in the lemon shark. However, recent studies by us (13, 15) on the effects of environmental dilution on the reabsorptive transport of urea in the renal tubules of the skate, *Raja erinacea*, have shown that reduction in the external salinity of the medium brings about a significant decrease in the active transport of urea by the renal tubules. In skates maintained in 100% seawater 94 ± 2.4% (mean ± S.E. of five fish) of the urea filtered at the glomerulus is reabsorbed by the renal tubules. In contrast only 66 ± 6.6% (four fish) of the filtered urea is reabsorbed by the kidneys of skates maintained in 50% seawater. Thus environmental dilution leads to an inhibition of the transport of urea in the renal tubules of elasmobranchs.

There are few completely freshwater elasmobranchs. They are all stingrays and belong to the family Potamotrygonidae. *Potamotrygon*, the most well known genus of the family, inhabits the upper Amazon River 4,000 - 5,000 kilometers from the sea. This elasmobranch has very low concentrations of urea in its body fluids, less than 1/300 the concentration found in marine elasmobranchs (1 vs. 350 - 400 μmoles/ml) (35). The relative absence of urea in *Potamotrygon* raises some interesting questions concerning possible biochemical differences between these freshwater rays and marine elasmobranchs. Is the low concentration of urea in the body fluids due to a low level of urea biosynthesis or an increased rate of urea excretion, or both? Assay of the activities of the ornithine-urea cycle enzymes in the liver of *Potamotrygon* showed (16) that the levels of these enzymes are all significantly lower than those in the livers of marine stingrays (Table 1). The level of the rate-limiting enzyme of the cycle, carbamoyl phosphate synthetase, in *Potamotrygon* is 1/10 - 1/20 the level of the enzyme in marine rays. In addition the rate of incorporation of ^{14}C-bicarbonate into urea by slices of liver from *Potamotrygon* is 1/100 that reported (28) for marine elasmobranchs (2.1 X 10^{-2} vs. 3.0 μmoles/g / hr). Marine elasmobranchs excrete only 1 - 2% of their total urea per day (Fig. 2). In contrast studies

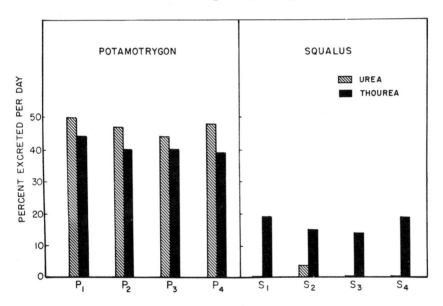

Fig. 2. Rate of excretion of ^{14}C-urea and ^{14}C-thiourea by the freshwater stingray *Potamotrygon* and the marine dogfish *Aqualus acanthias*. From Goldstein and Forster (16).

with ^{14}C-urea showed that *Potamotrygon* excrete approximately 50% of their total urea per day. In a study (16) comparing the rate of excretion of ^{14}C-thiourea, the sulfur analogue of urea which is not reabsorbed by the renal tubules of the elasmobranch kidney (29, 32),

Table 1. *Relation of ornithine-urea cycle activities to habitat in stingrays and lungfish*

Group	Species	Habitat	Enzyme Activities (μmoles/g liver / hr)			
			CPS	OCT	ASS	Arginase
Stingrays	*Dasyatis americana*	Marine	6.5 ± 1.3 (4)	14,360 ± 1,420 (6)	21.6 ± 2.1 (6)	34,880 ± 820 (6)
	Urolophus jamaicensis	Marine	4.5 (1)	8,540 (1)	16.5 (1)	13,920 (1)
	Potamotrygon sp.	Freshwater	0.36 ± .05 (4)	1,600 ± 210 (4)	9.4 ± .5 (4)	4,310 ± 240 (4)
Lungfish	*Protopterus aethiopicus*	Semiaquatic	31.2 (3)	1,675 (3)	6.6 (3)	34,800 (3)
	Neoceratodus forsteri	Aquatic	0.38 (2)	155 (2)	2.2 (2)	1,290 (2)

CPS = carbamoyl phosphate synthetase. OCT = ornithine carbamoyl transferase. ASS = arginine synthetase system. Values shown are means S.E. with numbers of fish in parentheses. From Goldstein, Janssens and Forster (17) and Goldstein and Forster (16).

we found that the two compounds were excreted at approximately the same rate (Fig. 2), indicating that urea is not actively reabsorbed by the renal tubules of *Potamotrygon*. Thus, adaptation to a permanent freshwater existence by this stingray is accompanied by reduction in the rate of urea biosynthesis and decreased tubular reabsorption of urea. It may be noted that the urea biosynthetic pathway is not entirely absent in *Potamotrygon* liver and perhaps some level of renal tubular transport may still exist. Future experiments to determine whether the low levels of both systems might be elevated by acclimatizing the fish to increasing environmental salinities seem worthwhile, since such experiments may shed light on environmental factors regulating the biosynthesis and active cellular transport of urea.

There are several fishes that can and do stay out of water for varying periods of time. The best studied among these is the African lungfish, *Protopterus* (19, 30). There are two surviving families of lungfishes, the *Lepidosirenidae*, which include the African *Protopterus* and the South American *Lepidosiren*; and the *Cerodontidae* represented by one surviving member, the Australian lungfish, *Neoceratodus forsteri*. *Lepidosiren* and *Protopterus* must breathe air at frequent intervals and they estivate in burrows out of water during dry seasons. In contrast, *Neoceratodus* appears to use its lung only as an accessory respiratory organ during heightened physical activity, and it cannot survive water deprivation by estivating. *Protopterus* and *Neoceratodus* are both ammonotelic in water (17). However, during estivation *Protopterus* converts the toxic end-product ammonia, which cannot be excreted, into urea (19, 30). We (17), therefore, compared the activities of the ornithine-urea cycle enzymes in the liver of *Protopterus* with those in *Neoceratodus* which never leaves the water and has no need to detoxify ammonia. The levels of all enzymes were significantly lower in *Neoceratodus* than in *Protopterus* (Table 1). The activity of carbamoyl phosphate synthetase, the rate-limiting enzyme of the cycle in the Australian lungfish was about 1/100th the level in the African lungfish. The rate of incorporation of ^{14}C-bicarbonate by liver slices was similarly reduced in *Neoceratodus* (17).

The early studies of Homer Smith (30) on the estivating lungfish showed that *Protopterus* was capable of converting large quantities of ammonia into urea during estivation; urea could comprise as much as 1% of the body weight at the end of one year of estivation. It was, therefore, of interest to determine what the route of urea formation was in this fish and the nature of the adaptation in nitrogen metabolism that enabled ammonia to be converted into urea during estivation. Our studies on isolated liver slices (11) and those of Janssens and Cohen on tissue homogenates (20) indicated that the ornithine-urea cycle was the major route of urea formation in *Protopterus*.

We found that there was no increase in the rate of incorporation of ^{14}C-bicarbonate into urea in liver slices taken from estivating lungfish (Table 2). Similarly, Janssens and Cohen observed that CPS was not elevated in the liver of estivating lungfish. In fact both the rate of ^{14}C-bicarbonate incorporation into urea and

Table 2. *Ornithine-urea cycle activity in estivating and aquatic African lungfish*, Protopterus sp.

Group	Hepatic Urea Synthesis[a] (μmoles/g / hr)	Enzyme Activities[b] (μmoles/mg protein / hr)	
		CPS	ASS
Aquatic	0.09	0.38	0.62
	1.3		
	2.4		
Estivating	0.04	0.20	0.47
	0.31		

CPS = carbamoyl phosphate synthetase. ASS = argininosuccinate synthetase.

[a]Data from Forster and Goldstein (11).

[b]Data from Janssens and Cohen (21).

the levels of CPS and argininosuccinate synthetase showed tendencies to be reduced during estivation (Table 2). It appears, therefore, that the adaptation in ammonia metabolism during estivation is not due to an increased capacity of the ornithine-urea cycle. Various suggestions have been put forward to explain the switch from ammonotelism to ureotelism during estivation, but non seems entirely adequate. Smith (30) noted that oxygen consumption was markedly reduced in estivating *Protopterus*. It may be that the ornithine-urea cycle continues to operate at the same level in estivating as in aquatic fish but in the former ammonia formed at a slower rate and, unable to be excreted, is totally incorporated in the ornithine-urea cycle and slowly accumulates as the non-toxic product, urea.

The South African toad, *Xenopus laevis* is strictly aquatic; it never leaves the water, but sometimes the water leaves it. During the dry periods when the swamps that it inhabits become dehydrated, *Xenopus* burrows into the mud and remains there until the water supply is restored. In water *Xenopus* is ammonotelic; it excretes 70 - 80% of its waste nitrogen as ammonia and 20 - 30% as urea (3). The major fraction of this nitrogen is excreted via the kidneys. During estivation renal function is presumably reduced to a low level and there is probably minimal excretion of nitrogenous end-products. However, under these conditions little ammonia accumulates in the body fluids but urea may increase to levels 15 times those found in aquatic toads (5). Thus, the need to detoxify ammonia during estivation is met by incorporating the toxic base into urea. The switch from ammonotelism to ureotelism seen during estivation in *Xenopus* may be artificially produced in the laboratory by keeping the animals in moist peat or in hyperosmotic saline (4). The two methods produce essentially the same condition, a relative lack of "free" water. Both methods have been used to study the mechanisms involved in the adaptation of nitrogen metabolism to water deprivation in *Xenopus*. We (24 - 26) have chosen to use hyperosmotic saline as the inducing medium since it is technically easier to study the toads in an aqueous environment then buried in peat. When the toads are first placed in hyperosmotic saline there is a marked reduction in urine flow (25) and in the excretion of ammonia and urea (Fig. 3). Later as urine flow returns toward (but remaining below) the freshwater level, excretion of nitrogenous end-products resumes. However urea becomes the predominant end-product, its excretion rate rising to levels 5 - 10 times that of freshwater toads,

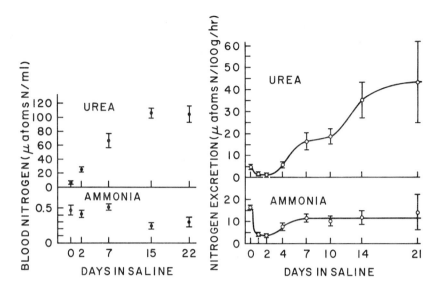

Fig. 3. Blood nitrogen levels and nitrogen excretion in *Xenopus laevis* during adaptation to hyperosmotic saline. Adapted from McBean and Goldstein (25).

whereas ammonia excretion remains at a level somewhat below that in freshwater (Fig. 3). In addition to the elevated level of urea excretion, there is an increase in the blood level of this nitrogenous end-product (but not ammonia) (Fig. 3) indicating that the biosynthesis of urea must be increased in toads maintained in hyperosmotic saline. The elevation in urea synthesis during osmotic stress is beneficial for two reasons. First it facilitates the detoxification of ammonia which cannot be readily excreted while renal function is reduced. Second, the accumulation of urea in the body fluids allows the toad to achieve osmotic equilibrium and survive in an otherwise hyperosmotic environment. The latter phenomenon is seen also in the adaptation of several other amphibia to saline media (see McBean and Goldstein (25, 26)).

We (25) have recently investigated the relative contribution of changes in urea biosynthesis to the increases in blood urea concentration during the adaptation of *Xenopus* to saline. An equation was derived from which plasma urea level could be calculated as a function of both urine flow and rate of urea biosynthesis. Since the urine of *Xenopus* has the same concentration of urea as the blood, the rate of change of blood urea concentration may be expressed as:

$$\frac{dP}{dt}(t) = \frac{A(t)}{S} - \frac{V(t)}{S} P(t)$$

where P = blood urea concentration, A = rate of urea biosynthesis, V = urine flow, S = urea space. The rate of urea biosynthesis was estimated indirectly by adding the rate of urea excretion at a given time to the rate of change in body urea content at that time. The latter was taken to be the product of the urea space and the rate of change in blood urea concentration. On the basis of data from other species, the urea space was assumed to be 60% of the body weight. Thus, starting with a known value of P at t = 0, then entering the successive experimental values for A, and integrating numerically (by computer), one can estimate the blood urea level at various times after transfer of toads to saline. By holding constant (at the freshwater level) the value for urine flow it is possible to observe the contribution of changes in urea biosynthesis to the elevation in blood urea concentration. The results of these computations (Fig. 4) indicate that changes in urea biosynthesis alone (Curve A) contribute only a fraction to the observed

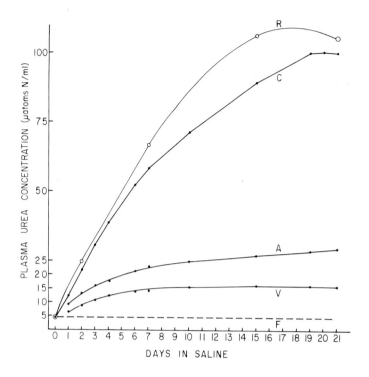

Fig. 4. Relationships between plasma urea concentration and changes in urine flow and urea biosynthesis in *Xenopus*. Computed theoretical effects of variation in urine flow (Curve V), synthesis (Curve A) and synthesis and urine flow (Curve C) on plasma urea levels are compared to observed changes (Curve R) in plasma urea concentration during adaptation of *Xenopus* to hyperosmotic saline. Line F shows freshwater plasma urea levels. From McBean and Goldstein (25).

rise in blood urea concentration (Curve R) (as do changes in urine flow (Curve V)) and that changes in urea biosynthesis and urine flow (Curve C) act synergistically to raise blood urea concentration.

Thus, both increases in urea biosynthesis and reduction in urine flow are needed to raise and maintain elevated blood urea concentrations.

We (26) have investigated several factors that might have been responsible for the rise in urea biosynthesis during osmotic stress in *Xenopus*. The rate of incorporation of ^{14}C-bicarbonate into urea by liver slices was elevated two-fold in toads maintained in hyperosmotic saline for 2 - 3 weeks. The activity of carbamoyl phosphate synthetase activity was also elevated four-fold by similar treatment. Janssens and Cohen (21) found that other enzymes of the ornithine-urea cycle: argininosuccinate synthetase, argininosuccinate lyase and arginase, were increased two-fold by maintaining *Xenopus* in hyperosmotic saline for 2 weeks. Thus, osmotic stress induces a general increase in enzymes of the ornithine-urea cycle. Since the response of CPS (as well as the other enzymes) to osmotic stress in *Xenopus* is rather unique, we (26) sought to determine whether the rise in enzyme activity following saline treatment was due to an increased concentration of enzyme protein or the result of conversion of preexisting inactive enzyme to a more active form. The method used to solve this problem was the antibody-titration technique of Feigelson and Greengard (9). In this technique, successively increasing amounts of enzyme activity (in tissue extract) from treated or untreated animals are added to a constant amount of enzyme antibody and the amount of enzyme activity remaining after incubation is determined. If the amount of enzyme activity required to titrate the antibody to neutrality (indicated by the appearance of enzyme activity in the assay mixture after incubation with the antibody) is the same for enzyme from both groups of animals then the treatment has not altered the ratio of active to inactive enzyme. We found that the amount of CPS antigen required to titrate CPS antibody to the equivalence point was the same for enzyme extracts from both freshwater and saline-treated toads (Fig. 5). This result

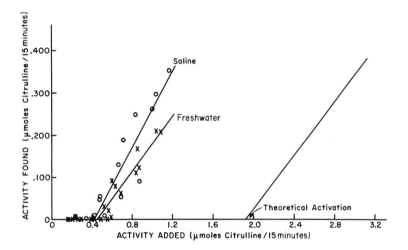

Fig. 5. Immunochemical analysis of the increase in hepatic carbamoyl phosphate synthetase level during adaptation of *Xenopus* to hyperosmotic saline. From McBean and Goldstein (26).

indicates that saline treatment does not alter the ratio of inactive to active enzyme in the liver of *Xenopus* and that the rise in CPS activity during osmotic stress is probably due to an increase in enzyme concentration.

The rise in levels of ornithine-urea cycle enzymes can not account for the rise in urea synthesis observed in *Xenopus* during osmotic stress. First, the two-fold rise in bicarbonate incorporation into urea is quantitatively insufficient to account for the fivefold rise in urea biosynthesis. Second, the most marked rise in urea biosynthesis occurs prior to the rise in carbamoyl phosphate synthetase activity (Fig. 6). A significant rise in CPS activity

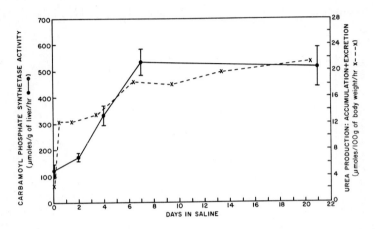

Fig. 6. Relation of hepatic level of carbamoyl phosphate synthetase to urea production in *Xenopus* during adaptation to hyperosmotic saline. From McBean and Goldstein (25).

was not detectable until 4 days after transfer of the toads to saline, while urea production increased more than four-fold during the first day of adaptation. However, the secondary rise in urea production observed between 3 to 7 days in saline may have been related to the elevated levels of CPS.

Since changes in levels of ornithine-urea cycle enzymes could not explain the increased rate of urea biosynthesis in *Xenopus* during osmotic stress, we examined other factors related to urea production that might have been altered by saline treatment. Concentrations of intermediates of the ornithine-urea cycle were assayed in quick-froze livers of saline-treated *Xenopus* and compared to concentrations found in freshwater controls. Maintaining toads in hyperosmotic saline for 2 - 3 weeks produced no significant changes in concentrations of ammonia, bicarbonate, ornithine, citrulline, aspartate and arginine (Table 3). Thus, at least in the steady-state, urea biosynthesis is

Table 3. *Concentrations of ornithine-urea cycle substrates and intermediates in liver of* Xenopus laevis

	Freshwater	Saline
	μmoles/g wet weight	
Ammonia	3.9 ± 0.76 (7)	3.1 ± 0.74 (8)
Bicarbonate (Plasma)	22 ± 3.3 (4)	16 ± 1.7 (6)
Ornithine	1.4 (2)	1.2 (2)
Citrulline	0.15 (1)	0.17 (2)
Aspartate	1.1 (1)	1.6 (2)
Arginine	0.05 (1)	0.03 (2)
Urea (Plasma)	3.4 ± 0.60 (8)	52 ± 3.3 (8)

Amino acid values are concentrations in samples containing pooled tissue from 3 frogs; number of samples shown in parentheses. Other values are means ± standard errors of samples from individual frogs, with number of animals in parentheses. From McBean and Goldstein (26).

elevated with little or no change in levels of intermediates of the ornithine-urea cycle. It seemed possible that the later phase of the rise in urea biosynthesis might have been due to the elevated levels of ornithine-urea cycle enzymes and that the elevated hepatic-urea synthesis seen early in the adaptation might have been the result of increases in substrate levels that may not have persisted. Elevated ammonia concentrations seemed particularly likely to occur, since, at the earlier times the excretion of ammonia was much reduced. Indeed, 12 hours after transfer to saline blood ammonia was elevated three-fold and liver ammonia two-fold. However, it seems unlikely that the two-fold rise in hepatic ammonia concentration could account for the four-fold rise in urea production. Thus, the basic biochemical factors underlying the switch from ammonotelism to ureotelism in *Xenopus* maintained in hyperosmotic saline still remain in doubt.

The influence of water availability on nitrogen metabolism in aquatic organisms has attracted the attention of biologists for many years. The early studies in this area were confined to describing gross phenomena and seeking possible relationships to the development and survival of these organisms. The result of these endeavors is that we have accumulated a mass of data on the formation and excretion of nitrogenous end-products in a variety of animals living in environments of different degrees of hydration. In some cases we know the basic biochemical pathways involved in formation of these end-products and the changes that these pathways undergo when animals switch from one environment to another. However, what is lacking is an understanding of the basic control systems regulating the activities of these pathways and how these controls influence the formation of nitrogenous end-products, at a molecular level, under different environmental conditions. It is to this problem that we must now turn our attention.

SUMMARY

Urea biosynthesis is adaptive in aquatic organisms; the activity of the urea biosynthetic pathway increases in response to the threat of ammonia intoxication or dehydration. Water deprivation, either absolute (dehydration) or relative (osmotic stress) leads to a switch from ammonotelism to ureotelism. This is observed in the evolution of the ornithine-urea cycle (OUC) in certain fishes. The freshwater stingray, *Potamotrygon*, is ammonotelic, in contrast to marine elasmobranchs, and the activity of the OUC and its composite enzymes are markedly lower in the liver of this ray than in the livers of marine rays. Similarly, hepatic OUC activity and enzyme levels are much higher in the estivating African lungfish, *Protopterus*, than in the Australian lungfish, *Neoceratodus*, which never estivates. Adaptation of urea biosynthesis to water deprivation occurs during estivation and osmotic stress in the freshwater South African toad, *Xenopus laevis*. Placing *Xenopus* in a hyperosmotic environment induces a switch from ammonotelism to ureotelism within 1 day. OUC and carbamoyl phosphate synthetase (CPS) activities are not changed during the initial period in saline but OUC activity rises 2-fold and CPS activity increases 5-fold in toads kept in saline for 1 - 2 weeks. The elevated CPS activity is due to an increased concentration of antigenic protein.

REFERENCES

1. Baldwin, E. *Comp. Biochem. Physiol.* 1:24, 1960.
2. Baldwin, E. *An Introduction to Comparative Biochemistry* (Fourth Edition). London, Cambridge Univ. Press, 1964.

3. Balinsky, J.B. and E. Baldwin. *J. Exp. Biol. 38*:695, 1961.
4. Balinsky, J.B., M.M. Cragg and E. Baldwin. *Comp. Biochem. Physiol. 3*:236, 1961.
5. Balinsky, J.B., E.L. Choritz, C.G.L. Coe and G.S. van der Schans. *Comp. Biochem. Physiol. 22*:59, 1967.
6. Brown, G.W., Jr. *Taxonomic Biochemistry and Serology*, edited by C.A. Leone. New York, Ronald Press, 1964. p. 407.
7. Delaunay, H. *Biol. Rev. 6*:265, 1931.
8. Fanelli, G.M. and L. Goldstein. *Comp. Biochem. Physiol. 13*:193, 1964.
9. Feigelson, P. and O. Greengard. *J. Biol. Chem. 237*:3714, 1962.
10. Florkin, M. and G. Dûchateau. *Arch. Intern. Physiol. 53*:267, 1943.
11. Forster, R.P. and L. Goldstein. *Science 153*:1650, 1966.
12. Forster, R.P. and L. Goldstein. *Fish Physiology, Volume I*, edited by W.S. Hoar and D.J. Randall. New York, Academic Press, 1965. p. 313.
13. Goldstein, L. *Urea and the Kidney*, edited by B. Schmidt-Nielsen and D.W.S. Kerr. Amsterdam, Excerpta Medica Foundation, 1970. p. 243.
14. Goldstein, L. and R.P. Forster. *Comp. Biochem. Physiol. 14*:567, 1965.
15. Goldstein, L. and R.P. Forster. *Am. J. Physiol. 220*:742, 1971.
16. Goldstein, L. and R.P. Forster. *Comp. Biochem. Physiol. 39B*:415,
17. Goldstein, L., P. Janssens and R.P. Forster. *Science 157*:316, 1967.

18. Goldstein, L., W.W. Oppelt and T.H. Maren. *Am. J. Physiol. 215*: 1493, 1968.
19. Janssens, P.A. *Comp. Biochem. Physiol. 11*:105, 1964.
20. Janssens, P.A. and P.P. Cohen. *Science 152*:358, 1966.
21. Janssens, P.A. and P.P. Cohen. *Comp. Biochem. Physiol. 24*:887, 1968.
22. Kempton, R.T. *Biol. Bull. 104*:45, 1953.
23. Manderscheid, H. *Biochem. Z. 263*:245, 1933.
24. McBean, R.L. and L. Goldstein. *Science 157*:931, 1967.
25. McBean, R.L. and L. Goldstein. *Am. J. Physiol. 219*:1115, 1970.
26. McBean, R.L. and L. Goldstein. *Am. J. Physiol. 219*:1124, 1970.
27. Munro, A.F. *Biochem. J. 54*:29, 1953.
28. Schooler, J.M., L. Goldstein, S.C. Hartman and R.P. Forster. *Comp. Biochem. Physiol. 18*:271, 1966.
29. Schmidt-Nielsen, B. and L. Rabinowitz. *Science 146*:1587, 1964.
30. Smith, H.W. *J. Biol. Chem. 88*:97, 1930.
31. Smith, H.W. *Am. J. Physiol. 98*:279, 1931.
32. Smith, H.W. *Am. J. Physiol. 98*:296, 1931.
33. Smith, H.W. *Biol. Rev. 11*:49, 1936.
34. Smith, H.W. *From Fish to Philosopher* (Natural History Library edition). Garden City, Doubleday and Co., 1961.
35. Thorson, T.B., C.M. Cowan and D.E. Watson. *Science 158*:375, 1967.
36. Thorson, T.B. *Sharks, Skates and Rays*, edited by P.W. Gilbert, R.F. Mathewson and D.P. Rall. Baltimore, Johns Hopkins Univ. Press, 1967. p. 265.
37. Watts, D.C. and R.L. Watts. *Comp. Biochem. Physiol. 17*:785, 1966.

MECHANISMS OF UREA EXCRETION BY THE VERTEBRATE KIDNEY[1]

B. Schmidt-Nielsen

Department of Biology
Case Western Reserve University
Cleveland, Ohio 44106

As has been discussed in the previous papers, the conversion of waste nitrogen into urea serves two major functions in vertebrates, namely: 1) detoxification of ammonia, and 2) osmotic water retention. Dr. Goldstein discussed the mechanism for regulation of urea synthesis. In this paper I shall emphasize the role of urea in osmotic water retention and urea excretion mechanisms, which control the plasma urea level.

OSMOTIC WATER RETENTION

On the phylogenetic tree shown in Fig. 1, the vertebrates are arranged according to their major nitrogenous waste products. Three groups excrete urea primarily: 1) animals inhabiting a marine environment (This group encompasses sharks and rays, the rat fish, and the only living *C.oelacanth*, Latimeria. The marine frog, *Rana cancrivora*, may also be included in this group.); 2) freshwater forms adapted to tolerate arid conditions (this group includes the African

---

[1]Source of support, National Institutes of Health, Grant AM09975.

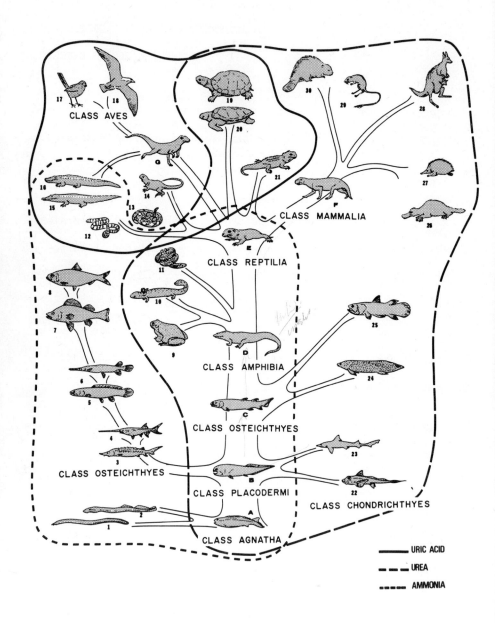

lung fish, most frogs and toads, and freshwater turtles); and 3) all mammals, regardless of habitat.

The plasma concentrations of urea and electrolytes in some of these animals are shown in Fig. 2. In the ureotelic marine fishes the Elasmobranchii, Holocephali and Coelocanth, the osmotic concentration of the plasma is higher than that of the marine environment (10). Urea plays a major osmotic role by constituting approximately one-third of the osmotically active substances in the plasma, with urea concentrations reaching 350 - 400 mM. In the crab-eating frog (*Rana crancrivora*) adapted to 80% sea water, the plasma urea concentration may reach values of 300 mM (4). In estivating African lung fish, *Protopterus*, the plasma urea concentration may reach values as high as 1,000 mM (23) and in the estivating spadefoot toad (*Scaphiopus hammondii*), urea concentrations up to 300 mM have been measured (20).

Dr. Goldstein emphasized the importance of urea synthesis to estivating animals for the detoxification of ammonia. Recently, however, investigations by Ruibal, Shoemaker and collaborators (11, 20) have shown that the role of osmotic water retention may be as important to estivating animals as to the marine forms. As shown in

---

Fig. 1 [p. 80]. Phylogeny and nitrogen excretion in the vertebrates. (From Schmidt-Nielsen and Mackay in *Clinical Disorders of Fluid and Electrolyte Metabolism*, in press.) The solid and dashed lines enclose groups of animals which utilize the three different forms of nitrogen as a major excretory product.

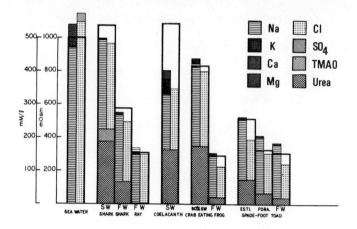

Fig. 2. Extracellular fluid composition of vertebrates which utilize urea in osmoregulation. (From Schmidt-Nielsen and Mackay in *Clinical Disorders of Fluid and Electrolyte Metabolism*, in press.) The dark line surrounding each bar gives the osmotic concentration. The composition of seawater is given for comparison. The concentrations of some of the constituents of the extracellular fluid are given within each bar. Osmolality, urea and trymethylamine oxide (TMAO) concentrations can be read directly from the mOs scale. Abbreviations below each bar indicate the environment to which the animal is adapted: adapted to seawater (SW); adapted to freshwater (FW); estivating in soil (ESTI); foraging in terrestrial habitat (FORA).

Fig. 3, Shoemaker *et al.* (20) found that the plasma urea concentration of estivating spadefoot toads (*Scaphiopus hammondii*) did not rise significantly during the period of estivation from September

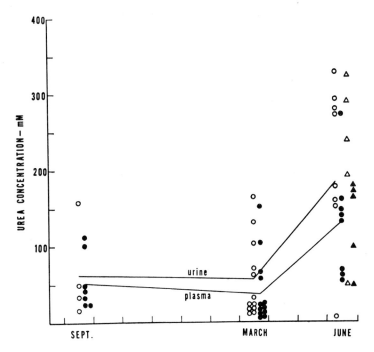

Fig. 3. Concentrations of urea in plasma (●) and urine (○) of *Scaphiopus hammondii* excavated at various seasons. Values for *S. couchii* in June (plasma, ▲; urine, △) shown for comparison. Lines fitted to mean values for *S. hammondii*. (From Shoemaker et al. (20).)

through March. However, between March and June the plasma urea concentration rose considerably. Samples of soil taken from the burrow sites were analyzed for water content. The equivalent osmotic concentration calculated from the mean soil moisture tension was about 60 mOs for September samples and 150 mOs for samples taken in March. However, by June the soil was much drier with an average moisture tension equivalent to a 600 mOs solution. The increase in plasma urea concentration raised the total osmotic concentration of the blood of the spadefoot toads above the osmotic concentration of the soil and favored the inward movement of water.

A similar interpretation of the ureotelism in reptilian embryos has been made by Packard (9). The eggs of the oviparous snake, *Coluber constrictor*, are deposited in moist places and the shell is water-permeable. The osmotic concentration inside the egg is raised through the production and accumulation of urea, and a favorable gradient for water is established. That water is taken up by the egg during the incubation period, is shown by the fact that the egg doubles in size prior to hatching. The embryo undergoes a biochemical metamorphosis from ureo- to uricotelism associated with the transition from embryonic to free living existence when osmotic water retention is no longer required.

Most adult reptiles and all birds are predominantly uricotelic, as shown in Fig. 1. The mammals are the only fully terrestrial animals which are completely ureotelic. In my opinion, the persistence of ureotelism in mammals is closely linked to the simultaneous development of the countercurrent kidney which must have been functioning by the time mammals became fully terrestrial. Urea accumulates in the renal medulla and increases its osmotic concentration, thereby favoring movement of water out of the collecting ducts. Therefore, the medullary urea accumulation is important for the concentration of the urine. In Fig. 4, cross sections of kidneys of various mammals are shown. The length of the renal medulla correlated with the ability to concentrate the urine (14). Chemical analysis of the various zones of the mammalian kidney has shown that urea accumulates

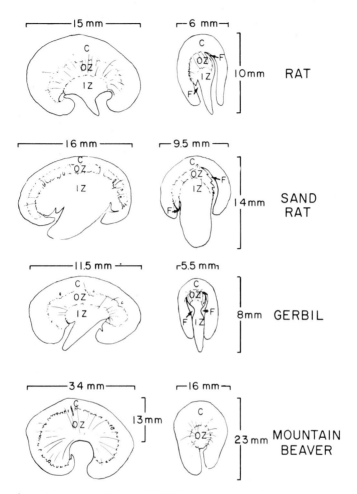

Fig. 4. Cross sections of kidneys of various rodents. Cortex (C), outer zone (OZ), inner zone (IZ). The sand rat and the gerbil are desert rodents with no inner zone. The mountain beaver is a water-loving animal with no inner zone of the medulla.

in the inner zone of the kidney in concentrations up to 1,000 mM (18) and constitutes about half of the osmotically active solutes that accumulate in the tissue (Fig. 5). Thus, urea serves as somewhat

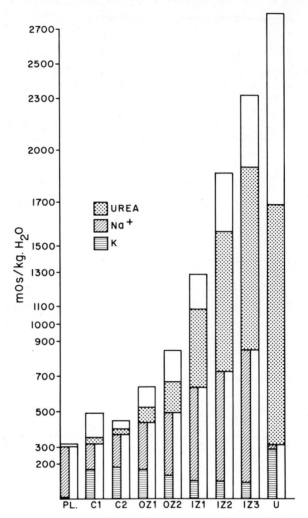

Fig. 5. The accumulation of urea and electrolytes from the kidney of a dog on a high protein diet.

similar osmotic function in the mammalian kidney as that found in estivating amphibians and some marine fishes.

Birds which are uricotelic also have a renal countercurrent mechanism and a modest concentrating ability. However, while in mammals all of the nephrons form loops of Henle and thus participate in the countercurrent system, only a small fraction of the bird nephrons have loops of Henle. Analyses of the renal cones (The equivalent of renal papillae) of chickens and turkeys (17) showed that urea constitutes less than 2% of the osmotically active solutes in the cone (Fig. 6). Thus, as could be expected, urea does not play

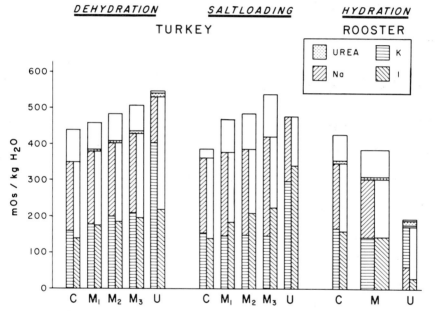

Fig. 6. Osmotic and ionic composition of the cortex and medulla of the kidney of turkey and rooster. (From Skadhauge and Schmidt-Nielsen [21].) Urea contributes only negligibly to the osmotic concentration of the medulla.

any role in the concentrating ability of the kidneys of the birds.

HOW IS THE PLASMA UREA CONCENTRATION DETERMINED?

From the foregoing, it can be seen that the plasma or tissue urea concentration may serve an important function in the water economy of a vertebrate. The level of the plasma urea concentration is determined by the rate of urea synthesis ($\dot{Q}$ urea synt. μM hr^{-1}) and the urea clearance (C$_{urea}$ ml hr^{-1}). A simple equation can be derived from the plasma urea concentration (P$_{urea}$ µM ml^{-1}) if the animal is in steady state; *i.e.*, rate of urea excretion ($\dot{Q}_{urea\ ex.}$) is equal to rate of urea synthesis. By definition:

$$C_{urea} = \frac{\dot{Q}_{urea\ ex.}}{P_{urea}} \tag{1}$$

then

$$P_{urea}(\mu M\ ml^{-1}) = \frac{\dot{Q}_{urea\ ex.}}{C_{urea}} = \frac{\dot{Q}_{urea\ synt.}}{C_{urea}} \tag{2}$$

If we keep $\dot{Q}_{urea\ ex.}$ constant we get a rectangular hyperbolic function which plotted on a log-log scale gives a straight line. By giving $\dot{Q}_{urea\ ex.}$ different values a series of parallel curves is obtained from which the plasma urea concentration, at steady state, can be estimated if we know the rate of urea excretion and the urea clearance (Fig. 7).

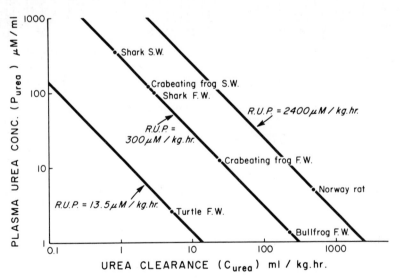

Fig. 7. The plasma urea concentration plotted against urea clearance according to the equation:

Urea excretion by the kidney

In the elasmobranch urea is excreted both by the gills and kidneys (22) while in amphibians and reptiles urea is eliminated entirely by the kidneys. Thus, in the latter the renal function determines to a large extent the plasma urea concentration. The renal urea clearance varies with a) the glomerular filtration rate (GFR), b) tubular secretion (T_s),   c) passive back diffusion (T_{pr}), and d) active reabsorption of urea (T_{ar}). Thus the amount of urea excreted is equal to the urine concentration (U_{urea}) times urine flow rate ($\dot{V}$):

$$U_{urea} \times \dot{V} = P_{urea} \times GFR + T_s - T_{pr} - T_{ar} \qquad (3)$$

In the following these mechanisms of regulation shall be discussed separately.

Glomerular filtration rate

Changes in GFR occur in most lower vertebrates with changes in salinity and with dehydration.

In an elasmobranch moving from a marine to a freshwater habitat, the GFR increases 4- to 5-fold.

In amphibians GFR decreases dramatically when the animal leaves the water and is exposed to a dry environment (13). Thus, in *Rana*

(Fig. 7 [p.88])

$$P_{urea} (\mu M\ ml^{-1}) = \frac{Q_{urea\ synt.}\ (\mu M\ hr^{-1})}{C_{urea}\ (ml\ kg^{-1}\ hr^{-1})}\ .$$

Values given have been estimated from data given in the literature (see text).

clamitans the filtration rate decreased from 30 ml kg^{-1} hr^{-1} when the animal was sitting in the water to 5 ml kg^{-1} hr^{-1} after three hours out of water (13). In the saltwater-adapted frog, *Rana crancrivora*, GFR fell from an average of 60 ml kg^{-1} hr^{-1} in freshwater, to 20 ml kg^{-1} hr^{-1} in 600 mOs saline water (19). In the turtle, *Pseudemys scripta*, which is predominantly ureotelic, the GFR decreased to 10 - 20% of normal when the animals were dehydrated for 12 - 40 hours (2). If the plasma osmotic concentration was increased by 20 mOs by intravenous administration of NaCl solution, the GFR decreased to zero. Conversely, in the predominantly uricotelic tortoise, *Gopherus agassizii*, no decrease in the GFR could be seen after dehydration, and only when the plasma osmolality had been raised by 100 mOs did the filtration decrease to zero (2). In other uricotelic reptiles dehydration has little effect upon the GFR (17).

Tubular secretion of urea

Active tubular secretion of urea into the renal tubules has been verified in frogs, but as I shall show in the following, it may be a much more general phenomenon among vertebrates. Forster (3) found that the urea secretion is characterized by a tubular maximum (T_m). Saturation of the transport mechanism is reached at a plasma urea concentration of 2 - 3 mM. The secretion of urea is blocked by 2,4-dinitrophenol. Forster showed that Probenecid also inhibits the urea secretion, presumably through a competitive mechanism. Marshall (8) first showed that the urea concentration of the kidneys of frogs is 6 - 8 times the urea concentration of the plasma.

This would indicate that the active transport of urea takes place across the basal membrane of the renal tubular epithelium and that the movement across the luminal membrane is passive. Another possibility was that the urea is not secreted across the basal membrane, but that urea is formed within the renal tubular cells. This hypothesis was supported by the finding that the renal tissue of frogs has a high concentration of arginase, while tadpoles which do not actively secrete urea show a low arginase concentration in the renal tissue. Studies with ^{14}C-labelled urea, however, proved the hypothesis not to be valid (15). From Fig. 8, it can be seen that when

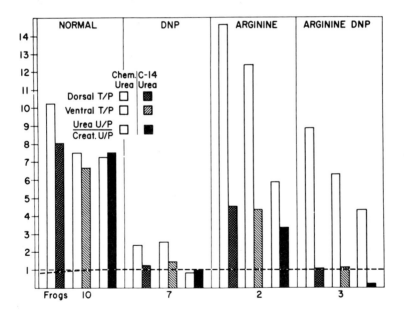

Fig. 8. Excretion and renal accumulation of urea by *Rana catesbeiana*. (From Schmidt-Nielsen and Shrauger [15].) Ratio for tissue/plasma urea concentration and urea/creatinine clearance are given on the ordinate.

^{14}C-urea was given to frogs, the chemical-urea urine/plasma concentration ratio was identical to the ^{14}C-urea urine/plasma ratio. Had a major part of the urea excreted been derived from *de novo* synthesis in the renal tubules, we would expect the chemical-urea clearance to exceed the ^{14}C-urea clearance. DNP blocked urea secretion and the chemical and the ^{14}C-urea clearances remained identical. However, when arginine was injected into the dorsal lymph sac of the frogs (Fig. 8), it served as a substrate for the renal arginase, and urea was formed in the tubular cells, thus enhancing the chemical-urea clearance, but not the ^{14}C-urea clearance. This shows that a large amount of the urea formed in the tubular cells during arginine infusion diffused across the luminal membrane and was excreted in the urine. Poisoning with DNP blocked tubular secretion of urea, but did not affect the arginine-induced formation of urea by the renal tubular cells as seen by the accumulation of unlabelled urea in the renal tissue and the chemical-urea clearance exceeding the filtration rate.

Administration of urea-related compounds such as methylurea, acetamide and thiourea showed that the molecules in which substitutions are made in the amino groups are not transported by the frog renal tubules, while thiourea is transported and accumulates in the renal tissue (Fig. 9). Furthermore, thiourea and urea showed competitive inhibition. The affinity of the transport system for thiourea is, however, only approximately one-tenth of that for thiourea (1

(Fig. 8 [p. 91]) White columns represent chemical urea, crosshatched columns ^{14}C-urea.

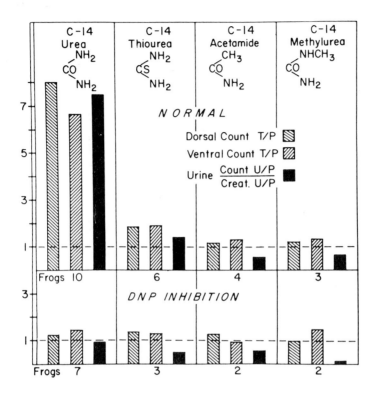

Fig. 9. Excretion and renal accumulation of urea and related compounds. (From Schmidt-Nielsen and Shrauger [15].) Results obtained after DNP administration are given in the lower part of the graph.

Passive back diffusion of urea

Micropuncture studies on the renal tubules of the frog, *Rana catesbeiana*, by Long (6) have shown that urea is secreted in the proximal and distal tubule (Fig. 10). Long's studies further revealed that in spite of the fact that water is reabsorbed from the fluid in the collecting ducts or in the bladder in this species there is no change in the urea clearance/inulin clearance in samples

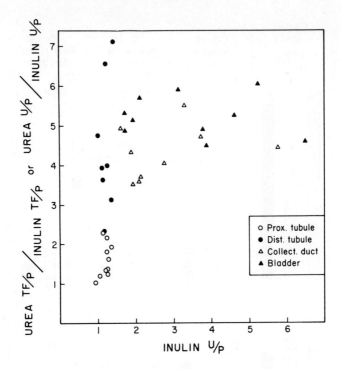

Fig. 10. Micropuncture data from *Rana catesbeiana*. (From W.S. Long, in preparation.) Tubular fluid over plasma urea concentration (TF/P) or urine/plasma (U/P) urea concentrations divided by the TF/P inulin concentration are plotted against inulin TF/P on the abscissa.

collected from these sites. This indicates a very low permeability to urea of the collecting duct or bladder wall.

In the frog, *Rana clamitans*, a similar low permeability to urea of the collecting duct, ureter, and bladder is indicated by the finding that the urea/creatinine clearance ratio is independent of tubular reabsorption of water (13). The urea/creatinine clearance ratio of bladder urine remained around 5 - 10 with tubular reabsorption of water varying from 50% to 95% of the filtrate (creatinine

U/P from 2 to 20). In contrast, semiaquatic frogs studied by Carlisky (1) showed a decrease in urea/inulin clearance ratio below unity with increasing water reabsorption (Fig. 11). The urea clearance exceeded

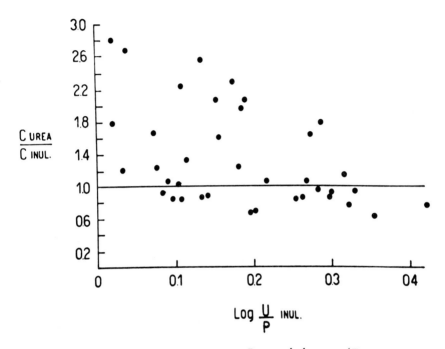

Fig. 11. Clearance studies on *Rana pipiens*. (From Carlisky [1].) The urea clearance (C_{urea}) divided by inulin (C_{inulin}) are plotted on the abscissa against the logarithms of the urine over the plasma inulin concentration. The fraction of filtered urea excreted decreases with increasing tubular reabsorption of water.

the filtration rate at high urine flows but with increasing water reabsorption the urea clearance decreased below the filtration rate. A similar phenomenon was observed in the saltwater frog, *Rana crancrivora* (19) (Fig. 12), and in the terrestrial toads, *Bufo marinus*

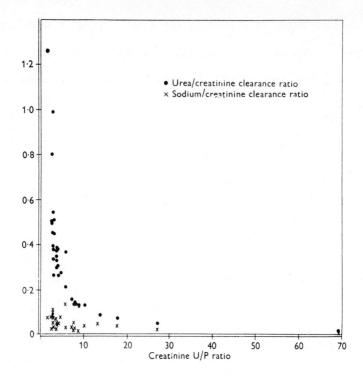

Fig. 12. Clearance studies in *Rana crancrivora*. (From Schmidt-Nielsen and Lee [19].) The urea over the creatinine clearance (–●–) is plotted against the creatinine U/P. At exposure to increasing salinity the fraction of filtered urea decreases.

and *Bufo alvarius* (Schmidt-Nielsen, unpublished), in which the passive back diffusion of urea increased with increasing water reabsorption. In all of these burrowing or saltwater-adapted species urea accumulation is of importance.

The difference in urea excretion in the various frogs has been referred to as a difference in active secretion (1). However, since the urea clearance in all of these cases exceeds the filtration rate at high urine flow and decreases with urine flow, Long (in preparation) pointed out that it indicates that urea is actively secreted in all of these amphibians, but that passive urea reabsorption is more pronounced in the more terrestrial forms. This hypothesis was supported by his finding that in *Rana pipiens* the renal tissue urea concentration was 3 to 4 times higher than the plasma urea concentration even when the urine flow was low and the urea clearance did not exceed the filtration rate.

Findings in some reptiles and even in birds of the urea clearance exceeding the GFR at high urine flows indicate that tubular urea secretion coupled with passive reabsorption is a rather widespread phenomenon. This has been found in the freshwater turtle (2) in *Sphenodon* (Schmidt-Nielsen, in preparation) and in the duck (24) (Fig. 13).

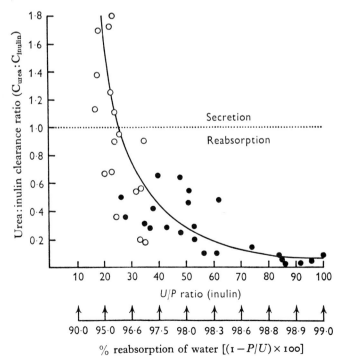

The passive reabsorption of urea is under the control of antidiuretic hormone (ADH) arginine vasotocin. Leaf and Hays (5) have shown that the toad bladder permeability to urea and some urea analogues increases greatly with ADH. Thus, dehydration during estivation or at exposure to high salinities causes an increased passive reabsorption of urea in toads and in some species of frogs. The same is true in mammals in which passive urea reabsorption in the collecting ducts is enhanced by ADH.

Tubular reabsorption of urea against a chemical concentration gradien

In the elasmobranchs the urine urea concentration is considerably lower than the plasma urea concentration (Table 1), indicating

Table 1. *Urine and plasma urea concentrations in various shark*

	Urine urea mM	Plasma urea mM
Squalus acanthias	93 ± 14	354 ± 3.6
Mustelus canis	202	285
Squatina dumerili	180	375
Heterodontus francisci	72	387
Rhinotriacis henlei	196	343

From Schmidt-Nielsen (12).

Fig. 13 (p. 97). Clearance studies from the duck (*Anas platyrhynchos*). (From Stewart *et al.* [24].) The urea over the inulin clearance is plotted against the inulin U/P ratio. Ducks maintained on freshwater (-O-) and on hypertonic saline (-●-).

active reabsorption of urea. Micropuncture studies showed that the reabsorption must take place in the distal tubule (16). In mammals on low protein intake urea is reabsorbed from the collecting duct against a gradient. However, the mechanism for the reabsorption of urea is peculiar in that it does not exhibit saturation of the transport mechanism, while the urea secretory mechanism in the amphibians shows all the characteristics of an active transport mechanism such as a low T_m, inhibition with metabolic poisons and competition, this is not the case for the reabsorptive mechanism. In the shark, Kempton (7) did not succeed in saturating the reabsorption of urea (Fig. 14) even at plasma urea concentrations as high as 750 mM.

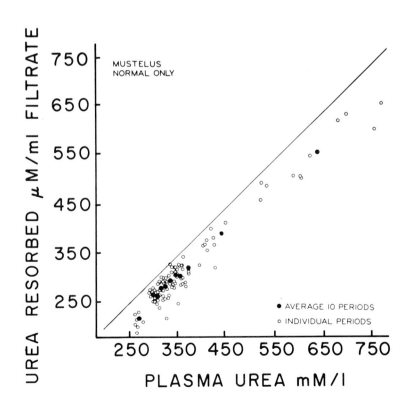

Urea analogs with substitutions in the amino group are reabsorbed by the tubules (Table 2) but thiourea is not reabsorbed.

Table 2. *Urine/plasma concentration in shark of urea and various urea analogues*

	Shark urine/plasma
Urea	0.41
Acetamide	0.76
Methylurea	0.62
Thiourea	2.74

From Schmidt-Nielsen (12).

This indicates a specificity of the reabsorptive mechanism that is different from that of the secretory mechanism found in the amphibians. The specificity is, however, similar to that observed in the toad bladder (5)

Raising the plasma acetamide concentration to 13.5 mM did not affect the urea reabsorption.

Recent anatomical studies of the shark renal tubules (Deetjen, *Bull. Mt. Desert Island Biol. Lab.*, in press) and Thurau (personal communication) have revealed an anatomical arrangement of the shark renal tubules that could conceivably cause passive urea movement out of the distal tubule.

Fig. 14 (p. 99). The relationship between reabsorption of urea per ml of filtrate and plasma urea concentration. (From Kempton [7].) The scale has been recalculated to mM.

CONCLUSION

The urea clearance and the synthesis of urea

From the foregoing it has been seen that the urea clearance is regulated by all the means possible as shown by Equation 3. In elasmobranchs it is regulated by variations in tubular reabsorption of urea and by variations in GFR. In amphibians it may be regulated only by variations in GFR as is the case in the most aquatic forms, or it may be regulated by variations in bladder and collecting duct permeability to urea (under the control of ADH). The latter mechanism is found in the more terrestrial forms and in burrowing or saline-adapted amphibians. In mammals the collecting duct permeability to urea is also under the control of ADH. Tubular urea secretion seems to be present in a wide variety of vertebrates but to what degree it may be regulated is not clear.

In Fig. 7 it is shown how the plasma urea concentration may be determined by variations in urea clearance or in urea synthesis. In the crab-eating frog the changes in plasma urea concentration reported by Gordon *et al.* (4) can easily be accounted for by the changes in GFR and passive back diffusion reported by Schmidt-Nielsen and Lee (19).

REFERENCES

1. Carlisky, J. In: *Urea and the Kidney*, edited by B. Schmidt-Nielsen and D.N.S. Kerr. Amsterdam: Excerpta Medica Foundation, 1970. p. 263.
2. Danzler, W.H. and B. Schmidt-Nielsen. *Amer. J. Physiol.* 210: 198, 1966.

3. Forster, R.P. *Am. J. Physiol. 179*:372, 1954.
4. Gordon, M.S., K. Schmidt-Nielsen and H.M. Kelly. *J. Exp. Biol. 38*:659, 1961.
5. Leaf, A. and R.M. Hays. *J. Gen Physiol. 45*:921, 1962.
6. Long, W.S. In: *Urea and the Kidney*, edited by B. Schmidt-Nielsen and D.N.S. Kerr. Amsterdam: Excerpta Medica Foundation, 1970. p. 216.
7. Kempton, Rudolf T. *Biol. Bull. 104*:45, 1953.
8. Marshall, E.K., Jr. *J. Cell. Comp. Physiol. 2*:349, 1932.
9. Packard, Gary C. *Am. Nat. 100*:667, 1966.
10. Read, L.J. In: *Urea and the Kidney*, edited by B. Schmidt-Nielsen and D.N.S. Kerr. Amsterdam: Excerpta Medica Foundation, 1970. p. 23.
11. Ruibal, R., W. Trevis and V. Roig. *Copeia 3*:571, 1969.
12. Schmidt-Nielsen, B. In: *Urea and the Kidney*, edited by B. Schmidt-Nielsen and D.N.S. Kerr. Amsterdam: Excerpta Medica Foundation, 1970. p. 252.
13. Schmidt-Nielsen, B. and R.P. Forster. *J. Cell. Comp. Physiol. 44*:233, 1954.
14. Schmidt-Nielsen, B. and R. O'Dell. *Am. J. Physiol. 200*:1119, 196
15. Schmidt-Nielsen, B. and C.R. Shrauger. *Am. J. Physiol. 205*:483, 1963.
16. Schmidt-Nielsen, B., K.J. Ullrich, G. Rumrich and W.S. Long. *Bull. Mt. Desert Island Biol. Lab. 6*:35, 1964.
17. Schmidt-Nielsen, B. and E. Skadhauge. *Am. J. Physiol. 212*:973, 1967.

18. Schmidt-Nielsen, B. and R.R. Robinson. *Am. J. Physiol. 218:* 1363, 1970.

19. Schmidt-Nielsen, Knut and Ping Lee. *J. Exp. Biol. 39:*167, 1962.

20. Shoemaker, V.H., L. McClanahan, R. and R. Ruibal. *Copeia 3:*585, 1969.

21. Skadhauge, E. and B. Schmidt-Nielsen. *Am. J. Physiol. 212:*793, 1967.

22. Smith, Homer W. *J. Biol. Chem. 81:*727, 1929.

23. Smith, Homer W. *J. Biol. Chem. 88:*97, 1930.

24. Stewart, D.J., W.N. Holmes and G. Fletcher. *J. Exp. Biol. 50:* 527, 1969.

25. Ullrich, K.J., G. Rumrich and B. Schmidt-Nielsen. *Pfl. Arch. ges. Physiol. 295:*147, 1967.

INTERACTION OF SALT AND AMMONIA

TRANSPORT IN AQUATIC ORGANISMS

J. Maetz

Groupe de Biologie Marine
Departement de Biologie
Commissariat a l'Energie Atomique
Station Zoologique
Villefranche/Mer 06
France

CONTRADICTORY VIEWS ON THE BIOLOGICAL SIGNIFICANCE OF AMMONOTELISM

The significance of the relationship between the principal end-products of products of protein and nucleic acid catabolism (ammonia, urea or uric acid) and osmoregulation was first recognized by Needham (63). He suggests that the predominant waste product is related to the needs of water economy during embryonic, larval and adult life. Ammonia having the lowest heat of combustion, is the most economic of these products. It is however the most toxic, but being highly soluble in water it is particularly fitting that the waste product should be ammonia if ample water is available for the rapid removal of this substance. A change of pattern of nitrogenous excretion is to be expected in animals colonizing areas where water availability is restricted, in order to permit water conservation. As life originated in water, Needham (63) followed later by Baldwin (4) suggested that the most primitive condition is "ammonotelism" while ureo- and uricotelism would have appeared later during the course of evolution.

Homer Smith (88) has vigorously challenged these concepts. He proposes that, at least in vertebrates, urea is the most primitive nitrogenous end-product and that ammonia production arises secondarily for the regulation of acid-base balance and for the conservation of cations. New weight has been given to this point of view by recent discoveries that the primitive Crossopterygian *Latimeria* is ureotelic (69) and that the teleosts which are undoubtedly ammonotelic have still retained the structural genes for the enzymes of the ornithine-urea cycle in their liver (37). Furthermore according to Smith, "the statement that ammonia is the most primitive and principal end-product is generally made without reference to *peripheral* formation of ammonia," outside the liver, in the effector organs of osmoregulation such as gills or kidney and, "ignores the abundant branchial excretion of ammonia, thus gravely distorting the facts." In fact, in Crustacea (64) in teleosts (27, 34, 65) and in aquatic larval amphibia (2, 19) the gills are the major sites of ammonia excretion while the skin intervenes in the aquatic adult forms of amphibians (5, 21).

Ammonia is diffusible across biological membranes. Its elimination into the external medium is therefore possible irrespective of the direction of the bulk flow of water across the membrane. This molecule does not need to be "flushed out" with water as is sometimes erroneously stated.

The gills of Crustacea and teleosts and the skin of Amphibia also play a major role in the maintenance of the mineral balance as demonstrated by the classical observations of Krogh (42) and Smith (87). Krogh was the first to surmise that at least in freshwater animals, the branchial or cutaneous excretion of ammonia may be related to cation absorption. More recently, the elucidation of mechanisms of ion uptake has made considerable progress with the help of radioactive tracers as a powerful tool of investigation. The determination of the unidirectional fluxes across membranes in relation to electrochemical gradient permitted a clear definition of active and passive transports (91, 92). In Crustacea (80 - 84), teleosts (28, 56, 59) and in Amphibia (39, 31) active transport of both ionic species Na^+ and Cl^- has been strongly established. Moreover, the existence of exchange mechanisms linking external Na^+ and Cl^- with the excretion of endogenous cation (NH_4^+) and anion (HCO_3^-) has also been demonstrated. A working model published by Maetz and Garcia Romeu (56) and which has been reproduced in several text books illustrates the possible mechanism of salt absorption in freshwater animals. Fig. 1 shows a slightly modified version suggesting tightly coupled exchanges situated on the external face of the epithelial cells.

In the first part of this contribution, we shall summarize the available evidence in favor of such a model. In the second part, we shall emphasize its inadequacy to fit all the facts and suggest alternative theories which should lead to further experimentation.

EVIDENCE FOR NH_4^+/Na^+ AND HCO_3^-/Cl^- EXCHANGE MECHANISMS UNDERLYING SALT ABSORPTION

Most of the observations were made on two freshwater species of

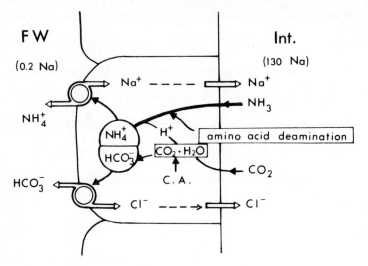

Fig. 1. Model for Na^+ and Cl^- absorption by the teleostean gill (modified from 56). A tight coupling between Na^+ uptake and NH_4^+ excretion and between Cl^- uptake and HCO_3^- release mediated by carriers located on the external face of the epithelial cells is suggested. CO_2 and NH_3 may cross the serosal membrane in the unionized form. They are excreted in the ionized form. Carbonic anhydrase (C.A.) plays a central role in catalyzing production of HCO_3^- and H^+, the proton being captured by NH_3. Part of NH_3 excreted results from enzymatic deamination of amino acids. In parentheses, the external and internal Na^+ concentrations are given.

animals: an invertebrate, *Astacus pallipes*, the crayfish (80 - 84) and a vertebrate, *Carassius auratus*, the goldfish (28, 42, 56). Similar evidence has been gained from the study of the larval amphibian *Ambystoma* (2, 19).

Independence of Na^+ and Cl^- absorption

In all the above mentioned species and also in both larval or adult frogs (3, 31, 75) more or less independent uptake of Na^+ and Cl^- from the external milieu is observed. For example when the animals are placed in dilute sodium chloride solutions (0.05 to 2 mM) sodium and chloride exchanges are frequently of very different intensities, resulting in net transfers which may be of the same sign, but of different values or they may be of opposite signs, indicating a net absorption of one ion and a loss of the other. While this independence is *occasionally* observed in animals kept in sodium chloride solutions or in running tap water, it is *constantly* seen in animals whose internal sodium or chloride level has been depleted by keeping them in artificial media lacking one or the other or both of these ions. Table 1 summarizes selected flux measurements made in the goldfish (28) which show most clearly this independence. Fig. 2 illustrates typical dissociations of Na^+ and Cl^- net fluxes obtained from "prepared" animals.

The independence of the absorption mechanisms is also demonstrated by the vigorous uptake of sodium or chloride ions notwithstanding the presence of a non-permeant counter-ion such as sulfate, choline or Mg^{++}. This phenomenon has first been mentioned by Krogh (42) in both freshwater invertebrates and vertebrates. For the goldfish, the absorption of either Na^+ or Cl^- appears to be unaffected by the nature of the counter-ion (28) but this is not the case in other animals. In *Astacus*, while sodium uptake seems to be independent of the nature of the counter-ion (81), the replacement of external

Table 1. *Independence of sodium and chloride absorption by the goldfish in sodium chloride solution*

Pretreatment solution	Na^+ exchange			Cl^- exchange		
	f_{in}*	f_{net}	f_{out}	f_{in}	f_{net}	f_{out}
Na_2SO_4	5	− 8	13	36	+ 9	27
	18	+ 4	14	73	+ 29	44
Choline chloride	57	+ 17	40	9	− 35	44
	40	+ 23	17	12	− 13	25
Deionized water	79	+ 34	45	21	− 44	65
	66	+ 48	18	−	− 31	−
NaCl	25	+ 2	23	31	+ 14	17
	40	+ 1	39	25	+ 16	9
	1	− 3	4	35	+ 18	17
	8	− 2	10	15	− 19	34
	27	+ 6	21	−	− 23	−

*f_{in}, influx; f_{net}, net flux; f_{out}, outflux in $\mu Eq \times h^{-1} \times 100\ g^{-1}$. Each horizontal line represents a separate experiment (28).

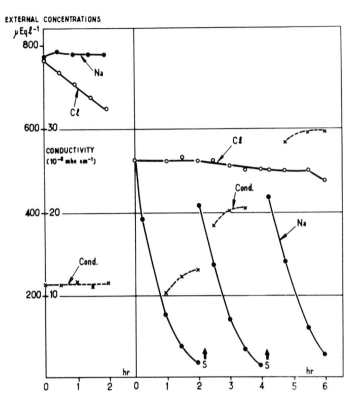

Fig. 2. Independence of Na^+ and Cl^- uptake from a NaCl solution by the goldfish (31). Variations of the external conductivity. Ordinates, external Na^+ or Cl^- concentrations or conductivity. Abscissae, time in hr. On the left, control fish (preadapted in NaCl solution). On the right, fish previously kept for 6 weeks in choline chloride solution. Arrows (S) indicate successive additions of Na_2SO_4 made to raise the external Na^+ concentration without altering that of Cl^-.

Na^+ by K^+ (impermeant) reduces in salt depleted animals the Cl^- net uptake (83). Chloride influx remains unaffected, but 90% of the unidirectional flux is accounted for by Cl^--Cl^- exchange (exchange-diffusion). Conversely in *Calyptocephallela* (a South American frog), chloride uptake is unaffected by the nature of the counter-ion, while the Na^+ net uptake is decreased by replacement of external Cl^- by SO_4^{2-} (31). Similar observations were made on the tadpole of *Rana catesbeiana* (3). These various observations clearly demonstrate, up to a certain degree, the independence between both uptake mechanisms. Shaw (84) has analyzed the nature of the feed-back mechanisms governing Na^+ and Cl^- balance and he suggests that two factors intervene 1) the absolute internal Na^+ or Cl^- level and 2) the relative Na^+/Cl^- concentrations possibly associated to changes of the plasma pH.

Evidence for separate cationic and anionic exchange processes

These differences in the absorption of ions of different charges can only be explained in conformity with the laws of electroneutrality of solutions, by exchange with endogenous ions of the same charge. The existence of such exchanges is confirmed by the results of conductivity measurements of the external medium. Fig. 2 illustrates such experiments showing that the conductivity remains constant or even increases slightly even though Na^+ or Cl^- were absorbed rapidly. Fig. 3 shows an experiment in which both ions are absorbed simultaneously and yet the conductivity increases. This figure also gives an example for the goldfish of the effects of a sudden decrease of the internal salt concentration upon the rates of salt absorption. This alteration is here produced by the intraperitoneal injection by means

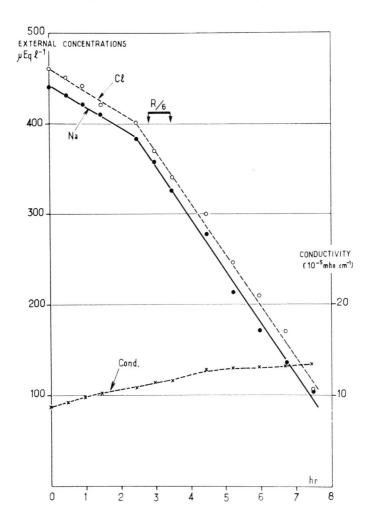

Fig. 3. Simultaneous absorption of Na^+ and Cl^- ions by the goldfish. Variations of the external conductivity (31). Coordinates as in the experiment on control fish in Fig. 2. The *Arrow R/6* indicates an intraperitoneal injection of hypotonic saline (20 mM) producing a simultaneous stimulation by a feed-back reaction of the Na^+ and Cl^- uptake rates.

of an indwelling catheter of a hypotonic saline solution. Such rapid feed-back mechanisms described by us (7) are similar to those reported by Shaw (84) discussed above.

Shaw (81) has made a more refined study of the conductivity changes in the external solution during salt uptake by the crayfish. He pointed out that stimulation of sodium uptake was followed by a diminution of the rate of increase of conductivity showing that uptake is not accompanied by the excretion of other metallic cations of equivalent mobility. In a second type of experiment, he compared the conductivity of external samples before and after passing through an ion-exchange column containing strongly basic resin which exchanges acid radicals with OH^- groups. If sodium were exchanged by the animal against non-metallic cations (H^+ or NH_4^+ ions for example) the resulting conductivity would be lowered after passing through the column. This is indeed what is observed.

On the other hand, metallic cations such as K^+ or Ca^{2+} may be eliminated at once as fulfilling an important role in *exchange* of Na^+, since their normal rate of loss is much lower than the maximal rate of sodium uptake (28, 81). In the case of Ca^{2+} a net absorption is sometimes observed (28, 16). Similarly, monovalent or divalent anions such as Br^-, I^- or SO_4^{2-} may be eliminated because of their relatively low concentration in the internal milieu or due to the impermeability of the gill to these ions (28). Moreover I^- is actively taken up by the teleostean gill (44). Finally, the only endogenous ions which may fulfill the role for the exchange are either NH_4^+ or H^+ for Na^+ and HCO_3^- or possibly OH^- for Cl^-.

In support of the postulated NH_4^+/Na^+ exchange is the fact that the gill or the skin are the principal site for nitrogenous excretion in ammonotelic animals as discussed above. Table 2 gives a summary of the measured rates of ammonia excretion across gills or skin in various vertebrates and invertebrates. Except for adult amphibians which have switched to ureotelism, the rates listed appears sufficient to account for the maximal rates of Na^+ uptake given for the same species and also recorded in Table 2. This point will be discussed again in the next section.

The sources of the ammonia excreted need to be considered as Smith suggested that, at least in teleosts, the major site of production is the effector of osmoregulation itself (86, 88). He considered that in this, the gill would play a role very similar to that played by the mammalian kidney in relation to acid-base balance regulation. Indirect evidence confirming Smith's view was found in studies that showed significant glutaminase and glutamic dehydrogenase in the branchial tissue (33). The maximal rate of ammonia production by gill homogenates was found sufficient to account for the ammonia excretion rate observed *in vivo* in *Myoxocephalus scorpius*. However, the activity of an enzyme as measured *in vitro* does not necessarily correspond with its performance *in situ*. Pequin's observations (65) on the carp showed that most if not all the ammonia excreted by the gill may be accounted for by the branchial clearance This was demonstrated by comparing the ammonia content of the blood entering and leaving the gill and taking into account the cardiac output. It was also noticed that the ammonia content was highest

Table 2. Comparative rates of ammonia excretion and sodium uptake in various freshwater animals. Rates in µEq or µM $\times$ h^{-1} $\times$ 100 g^{-1}

Species	Ammonia excretion			Sodium net uptake		
	t°	rate	ref.	t°	rate	ref.
Crustacea						
Astacus pallipes	12–13	38	(82)	12–13	43.5	(80,82)
Oronectes rusticus	22–25	92	(79)	–	–	–
Teleosts						
Salmo gairdnerii	13	22	(26)			
Anguilla anguilla	?	16	(47)			
Cyprinus carpio	7	17.5	(67)			
	20	48	(67)			
	19	100	(71)			
Carassius auratus	25	100	(16)	18–23	90	(28)
Amphibians (larval)						
Ambystoma tigrinum	15	7.1	(19)	15	13	(19)
Ambystoma gracile	15	25.0	(2)	15	15	(2)
Rana catesbeiana	20–24	35.0	(3)	20–24	50	(3)
Amphibians (adult)						
Calyptocephallela gayi	20	1.6	(31)	20	10.8	(31)
Leptodactylus ocellatus	20	0.9	(30)	20	3.7	(30)
Xenopus laevis	20	4	(5)			
Necturus maculatus	22–25	12.5	(21)			
Rana catesbeiana	20–24	7.0	(3)		– 1.0	(3)

in the blood leaving the liver, suggesting that it is the major source of ammonia. This was later confirmed (67) by perfusion studies with various precursors of ammonia production. The kidney also accounts for about one-third in this production, while the gill is considered as having a purely passive role in ammonia diffusion across the epithelium. To evaluate the relative importance of ammonia diffusion between the preformed waste-product and its formation *de novo* by the gill tissue, Goldstein and his colleagues (34) used the Fick's principle to measure the cardiac output while simultaneously determining the rate of ammonia excretion in *Myosocephalus in vivo*. Ammonia, amino acid and glutamine content of the blood entering and leaving the gills were measured. They concluded that about 40% of the ammonia is produced by deamination of amino acids, while 60% of the amount excreted is due to branchial clearance. Glutamine does not seem to be deaminated at all in the gills. As for the role of branchial glutamic acid dehydrogenase, Schoffeniels (78) claims that it plays a role in intracellular H^+ transport and that in the gills of Crustacea it favors the reductive amination of α-ketoglutaric acid rather than the reverse reaction. This occurs particularly when inorganic ions which are known to be inhibitors of the reverse reaction are present (12, 47). The whole problem of the source of excreted ammonia needs therefore to be reinvestigated. Other aspects of the problem of ammonia clearance by the gill, notably its passive nature, will be discussed in the next section.

The fact that bicarbonate ions may be reasonably assumed to be the endogenous ion to be exchanged against Cl^- will be discussed more briefly. The gill as a respiratory organ is undoubtedly in both fish and Crustacea the major route of CO_2 excretion. The role of the skin relative to the lungs in adult frogs is discussed by Garcia Romeu and his colleagues (31). The amounts of CO_2 excreted (up to 500 µmole X h^{-1} X 100 g^{-1}) is more than sufficient to cover the needs for the postulated HCO_3^-/Cl^- exchange. Furthermore in both fish and Crustacea, carbonic anhydrase, an enzyme which catalyzes the following reaction, is present in the gill tissue (43, 49 - 51):

$$CO_2 + H_2O \rightleftarrows H_2CO_3 \rightleftarrows HCO_3^- + H^+. \qquad (1)$$

For these reasons, the possibility of NH_4^+/Na^+ and HCO_3^-/Cl^- exchanges already suggested by Krogh (42) was subjected to experimental analysis (82, 56). Two methods of approach may be considered. The *direct method* involving physico-chemical analysis of the external medium during ion transport is problematic, because of the difficulty of determining the ratios of ionized and unionized forms of NH_3 and CO_2 crossing the membrane. This point will be discussed at length in the next section. The *indirect method* is easier to carry out. This consists of an analysis of the effects on sodium and chloride absorption rates of experimental changes of ammonia or bicarbonate concentrations in the external and internal media.

Indirect evidence for the involvement of NH_4^+ and HCO_3^-

1. Addition of ammonia or bicarbonate ions to the external medium.

The effect of both of these ions on ion balance in freshwater animals is of practical importance as Na^+ and Cl^- may form, in natural waters, only a small part of the total ionic content. Bicarbonate is often present in relatively high concentration in "hard" waters and ammonia may accumulate in confined volumes of water. In closed-circuit experiments necessary for determination of Na^+ and Cl^- uptake, both NH_4^+ and HCO_3^- may accumulate and reach relatively high concentrations (16, 71).

Shaw (82) was first to observe on *Astacus* that addition of NH_4^+ to the external medium produces an inhibition of the Na^+ influx and net uptake. The effect is directly proportional to the NH_4^+/Na^+ ratio of the concentrations. For a ratio of 20:1, the inhibition attains 80%. By increasing the external sodium concentration, the inhibition can be abolished. Inhibition is produced at very low concentration, lower in fact than the prevailing internal concentration (23). Furthermore, the inhibition is specific for sodium, as no effect on chloride transport is observed.

Similar observations were made on the goldfish (56). Addition of NH_4^+ inhibits Na^+ influx and induces a net sodium loss, the effect being reversible upon removal of the added salt by rinsing. Fig. 4 illustrates a typical experiment. The experiments made by us were however more crude than those reported by Shaw. NH_4^+/Na^+ concentration ratios up to 50 were used, the final concentration of ammonia being very high (up to 10 mM), certainly higher than the internal ammonia level. Furthermore no attempt was made to correlate the NH_4^+/Na^+ ratio with the importance of the inhibition. In the goldfish as in the crayfish, the inhibition is specific as no effect on Cl^- transfer is observed. Table 3 summarizes the data obtained in the goldfish. One point needs to be discussed in view of the

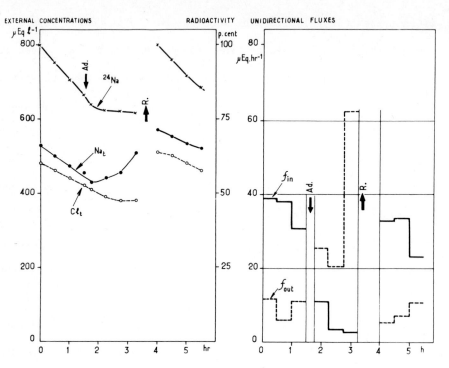

Fig. 4. Effect of addition of NH_4^+ to the external medium on the Na^+ exchange of the goldfish (56). On the left, ordinates, concentrations of ^{24}Na in % of initial radioactivity and of total Na^+ and Cl^- concentrations. On the right, influx and outflux of Na^+ for one-half hr periods in $\mu Eq \times hr^{-1}$ per fish (weight 85 gm). Abscissa, time in hr. *Arrow Ad* marks addition of a 5 ml of 1 M $(NH_4)_2SO_4$ to give a 9 mM concentration in the aquarium water. *Arrow R* marks the start of "rinsing."

possibility of H^+/Na^+ exchanges (see below). The added solutions of ammonium sulfate used by us were slightly acid, but the pH changes observed after addition were small, presumably because the acidity was buffered by the bicarbonate accumulated in the closed-circuit bath (see 16). In a series of unpublished experiments, we studied

Table 3. Effect of addition of ammonium ions into the external medium on the sodium exchanges of the goldfish (56).

	fin			fnet			fout		
Before	After	Dif.§	Before	After	Dif.	Before	After	Dif.

I. Addition of ammonium sulfate (n = 7)

| 34.1 | 16.0 | − 18.1** | + 13.5 | − 18.5 | − 32.0** | 20.6 | 34.5 | + 13.9* |
| ± 7.4 | ± 5.0 | ± 3.8 | ± 12.5 | ± 15.8 | ± 6.6 | ± 7.6 | ± 12.7 | ± 5.6 |

II. Addition of ammonium carbonate (n = 2)

| 44.6 | 8.5 | − 36.1 | + 16.1 | − 12.9 | − 39.0 | 28.5 | 22.5 | − 6.0 |

The means and standard error of the mean are given.

**P $\leq$ 0.01   *P $\leq$ 0.05 §Dif. = difference of paired data.

the effects of the addition of alkaline solutions of ammonium carbonate which were also effective in inhibiting Na^+ uptake. These experiments are included in Table 3. At the high concentrations of added ammonia, it is probable that ammonia excretion was totally blocked in the goldfish. This is suggested by experiments performed on the trout *Salmo gairdnerii* (27) in which the ammonia output is reduced by half with 0.5 mM ammonia in the bath. Similar observations were obtained on the carp (71).

The effects on Na^+ uptake and the reversal of the Na^+ net uptake were interpreted by us as resulting from the inhibition of NH_4^+ excretion and the reversal of the normal NH_4^+/Na^+ exchange. The possibility of an influx of ammonia added to the external medium remains to be confirmed however. Shaw (82) suggests that NH_4^+ ions are competing with Na^+ for the available transport sites, the affinity for ammonium exchange mechanism.

Similarly, addition of bicarbonate ions (in the form of potassium salt) into the external medium produces an inhibition of Cl^- uptake. This series of experiments showed conclusively that neither HCO_3^- nor K^+ ions added to the bath at concentrations 25 times than the prevailing Na^+ concentration cause an effect on the Na^+ uptake (see also 53). This observation illustrates the specificity of the observed action on Cl^- uptake and confirms further the specificity of the effects of NH_4^+ on Na^+ uptake. Again the effects on chloride absorption were shown to be reversible after rinsing the aquarium and adding fresh medium. The fully restored exchange pattern shows that no pathological effects or irreversible damage to the gill epithelium occurs with the high concentrations of added *ammonium* or bicarbonate used.

2. Injections of ammonium and bicarbonate ions into the internal medium.

Fig. 5 illustrates the effects of ammonium sulfate injection into the intraperitoneal cavity, without perturbing the fish.

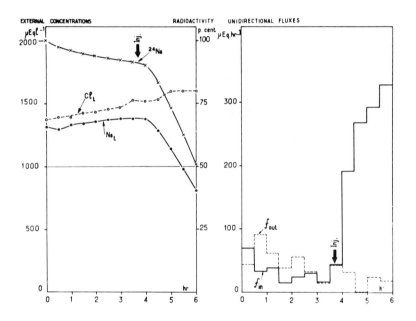

Fig. 5. Effect of intraperitoneal injection of NH_4^+ on the Na^+ exchange of the goldfish. Coordinates as in Fig. 3 (56). The *Arrow Inj* indicates the injection of 2 ml 1 M $(NH_4)_2SO_4$ equivalent to 1.25 µEq NH_3 per 100 g. (Weight of fish = 312 g)

Approximately 20 min after injection, there is a considerable increase in sodium influx and in the net flux, whereas the outflux remains unchanged. Table 4 presents data from experiments of this type. The quantity of ammonia injected varied from 800 to 1400 μmole X 100 g^{-1}. Control injections with sodium sulfate remain without effect.

Preliminary experiments on *Ameiurus nebulosus* (95) show that ammonium chloride injected intraperitoneally is excreted mainly across the gills within a few hours. In the carp also, a 100% increase of *ammonia* excretion is observed after intravenous injection of *ammonium* chloride administration (95).[1] This has been confirmed by us in collaboration with B. Lahlou in a preliminary experiment on the toadfish. The origin of this acidosis in animals with a notoriously low ornithine-urea cycle enzyme activity (37) remains to be determined. In a small series of unpublished experiments in collaboration with Garcia Romeu, injections of alkaline ammonium carbonate solutions into the intraperitoneal cavity (450 to 900 μmole X 100 g^{-1}) were performed, in order to obviate unspecific effects due to acidosis. Very similar effects on Na^+ influx and net uptake were obtained. These results are included in Table 4. Injection of glutamine into the carp is followed by rapid deamination by the liver and an increase in the level of circulating *ammonium* as well as an sugmentation of *ammonia* excretion (67). In a preliminary experiment, a similar treatment on the goldfish produced a prompt 300% increase of the sodium influx and a parallel augmentation of the net uptake (56).

---

[1] In the catfish, a mild acidosis is reported after ammonium chloride administration (95).

Table 4. *Effect of injection of ammonium ions into the internal medium on the sodium exchanges of the goldfish (56)*

	fin			fnet			fout		
Before	After	Dif.	Before	After	Dif.	Before	After	Dif.
I. Injection of ammonium sulfate (n = 7)									
25.4	52.1	+ 26.7**	- 14.3	+ 19.4	+ 33.1*	39.7	32.7	- 7.0
± 5.5	± 7.1	± 7.2	± 10.3	± 13.3	± 10.7	± 12.4	± 9.0	± 8.2
II. Injection of ammonium carbonate (n = 4)									
14.3	29.0	+ 14.7	- 8.5	+ 9.7	+ 18.2	22.8	19.3	- 3.5
± 8.5	± 17.0	± 8.6	± 1.9	± 10.3	± 11.6	± 9.6	± 7.4	± 3.2

See Table 3 for the meaning of the signs.

The effects of injected ammonia were even more spectacular in a series of experiments performed on the euryhaline European eel (29); 400 μmole X 100 g^{-1} of ammonium produced a five-fold increase of the Na^+ influx and a parallel increase of the net uptake. Similar observations were made on the flounder (59).

Although no attempt was made to correlate the extra amounts of Na^+ taken up with NH_4^+ excreted, Maetz and Garcia Romeu (56) and Garcia Romeu and Motais (29) conclude from these experiments that the results are consistant with the hypothesis of NH_4^+/Na^+ exchange.

Parallel experiments consisting of injections of bicarbonate ions were shown to produce promptly a specific increase of the chlorine influx and net uptake, while the outflux remained unchanged. Injection of acetazolamide a potent inhibitor of carbonic anhydrase is followed by a 75% decrease of the chloride uptake (56). These observations are in accord with the proposed HCO_3^-/Cl^- exchange model. Inhibition of carbonic anhydrase is followed by apparently "unspecific" effects because Na^+ uptake is also inhibited (51). The model proposed however (see Fig. 1) suggests that hydration of CO_2 which is accelerated by the enzyme also produces a proton which is captured by NH_3 producing NH_4^+, available for Na^+ exchange. Thus a "cross-effect" of the enzyme inhibition is to be expected. It would be of interest to verify whether ammonia excretion is altered by acetozolamide treatment.

Inclusion of the proposed exchanges into carrier-mediated active transports.

When ionic exchanges linked with uptake of ions across membranes from a medium of low sodium and chloride concentration are considered, the question arises whether the mechanisms involve endothermic processes either of the type "facilitated diffusion" or "active transport" (91, 92). All the available evidence points to active transport of both Na^+ and Cl^-, as suggested by Krogh (42). It is of particular interest in this respect that skins of amphibians *Rana esculenta* and *R. temporaria* gave contradictory results *in vivo* (39) and *in vitro* (91, 92). While active transport of Na^+ is observed in both types of experiments, transport of chloride against an electrochemical gradient is only observed *in vivo*. Similar observations were obtained from skin of the tadpole and adult form of *Rana catesbeiana* (3).

In salt-depleted goldfish, the ratio of influx *vs.* outflux (f_{in}/f_{out}) may be as high as 2.5 for Cl^- and 5.5 for Na^+ for a ratio of internal to external concentrations and presumably activities (C_{int}/C_{ext}) averaging 300 for Na^+ and 650 for Cl^- (56). Applying Ussing's criterion to calculate the difference of potential that would explain these ratios in terms of passive transport, a difference of 180 - 190 mV positive inside for chloride and negative inside for sodium should prevail. Such a high bioelectrical potential has never been encountered. Similar calculations were made for the crayfish (83, 11) and for the larval salamander and bullfrog (23, 19) suggesting active transport of both ionic species.

The few reports of the potential differences occurring across the boundaries of freshwater animals are characterized by sharp contrast. On the one hand, the crayfish and freshwater teleosteans and larval bullfrog seem to exhibit difference of potential negative inside (3, 11, 36, 55, 86) although contradictory reports have appeared (10). It would be of interest to verify whether in these species Cl^- transport is electrogenic. Unfortunately in the eel, for which potentials as high as 20 mV have been reported, chloride transport is nil (29). On the other hand, for larval *Ambystoma* and all adult amphibians investigated a difference of potential positive inside has been reported (1 - 3, 9, 19). Admittedly sodium transport plays a central role in electrogenesis in all these species, especially as potential differences are clearly related to variations of external sodium concentrations.

In the crayfish (80, 83, 84) as in all the Crustacea so far investigated (84, 46), the sodium and chloride influx increase as a function of external Na^+ or Cl^- concentrations respectively. All the curves obtained by plotting influx against concentration show a pattern of saturation kinetics and they fit with the well known Michaelis-Menton equation:

$$f_{in} = \frac{F_{max} \times C_{ext}}{K_m + C_{ext}} \qquad (2)$$

where f_{in} is the Na^+ or Cl^- influx; f_{max}, the maximal rate of influx; K_m, the "Michaelis-Menton constant", that is the concentration permitting half maximal influx; C_{ext}, the external concentration of Na^+ or Cl^-. In the case of the crayfish, the outflux of both ionic species also shows some variation in relation to external Na^+ or Cl^- concentrations. Circumstantial evidence points to the possibility of subdividing the outflux into two components, the *passive outflux* as measured in animals kept in deionized water, and an *exchange-diffusion* component which is external Na^+ or Cl^- concentration dependent (80 - 83). This component which corresponds to a Na^+/Na^+ and Cl^-/Cl^- exchange is seen only as a result of the use of isotopic tracers. Its net effect on ionic transport being nil, it does not intervene in the net uptake, while it may represent part of the ionic influx (92).

In the flounder adapted for at least three weeks to freshwater, the influx of Na^+ was also determined as a function of external sodium concentration. Fig. 6 illustrates the curve at external concentrations below 10 mM (57). The lower part of the figure represents

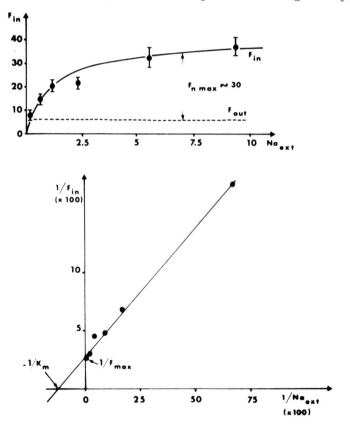

Fig. 6. Relationship between Na^+ influx and external Na^+ concentration in the freshwater flounder (57). The Na^+

the reciprocal plot which permitted evaluation of the parameters: f_{max} = 35 μmole X h^{-1} X 100 g^{-1} and K_m = 0.8 mM.

At concentrations above 10 mM a more complex relationship is observed between f_{in} and Na^+_{ext} (53, 54) as a passive influx component which is directly proportional to Na^+_{ext} intervenes.

Also shown on Fig. 5 is the sodium (branchial) outflux recently evaluated (48) and which shows only small variations upon a wide range of external salinity changes (see 59). Obviously the exchange diffusion-component which is of the utmost importance in the seawater adapted flounder (59) plays no role in the freshwater adapted form. Chloride flux has not been studied in relation to Cl^-_{ext}.

Similar curves relating Na^+ influx or net uptake vs. [Na^+_{ext}] have been published for amphibians whether in larval (1 - 3) or in adult forms (3, 9, 41) *in vivo* and *in vitro*.

All the observed facts which fit with Michaelis-Menton kinetics suggest by analogy to the concept of enzyme-substrate combination, the existence of carriers which are saturable by either Na^+ and Cl^- and which may show a certain amount of "inefficiency" in the form of exchange-diffusion. As suggested by Fig. 1 the hypothetical carriers would be not only responsible for Na^+ and Cl^- transport but

---

(Fig. 6 [p. 129]) outflux is also shown, as well as the maximal net uptake. Fluxes in μEq X h^{-1} X 100 g^{-1}. Concentrations in mEq per liter. The mean flux and standard error of the mean (n = 10) are given. Below, the reciprocal plot is represented permitting evaluation of the parameters f_{max} and K_m.

also for NH_4^+ and HCO_3^- extrusion, both absorption and excretion being tightly coupled. This obligatory exchange would be comparable to the Na^+/K^+ exchange pump which has been shown to occur in many biological membranes (85). This model takes into account the presence in the branchial cell of deaminating enzymes which are sources of NH_3 and of carbonic anhydrase. This enzyme is responsible for the production of HCO_3^- and H^+ which is presumably captured by NH_3 to form NH_4^+. The mobile carriers are supposed to be situated on the external membrane of the epithelial cell. If they were located on the serosal side, Na^+ entering the cell would compete for the carrier with NH_4^+, and Cl^- would compete for HCO_3^-.

The next section will be a critical evaluation of this model involving *obligatory* exchanges.

DEGREE OF COUPLING IN NH_4^+/Na^+ AND HCO_3^-/Cl^- EXCHANGES. DIRECT EVIDENCE

As has already been emphasized, the experiments described above are based on *indirect evidence*. If there is an obligatory exchange between NH_4^+ and Na^+ or HCO_3^- and Cl^-, a stoichiometric relationship between the rates of excretion and absorption should be observed for both couples of ionic species.

Absence of stoichiometric relationship between NH_3 excretion and Na^+ uptake: necessity of an alternative model.

Because of the difficulties encountered which have been briefly mentioned above, very few studies have appeared attempting to demonstrate directly an obligatory exchange.

Fig. 7 illustrates two such attempts: one concerning the crayfish (81) represented in the left hand side, and one related to the

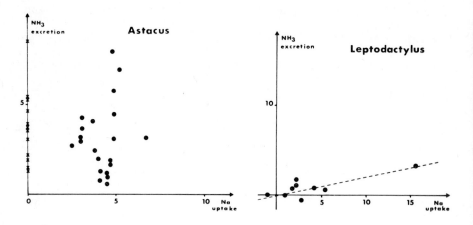

Fig. 7. Correlation between Na^+ uptake and NH_4^+ production (according to 82, 30). Ordinate, NH_4^+ branchial excretion in $\mu Eq \times h^{-1} \times 100\ g^{-1}$. Abscissa, Na^+ uptake in the same units. No correlation is observed in the crayfish (left), while a good correlation exists for the frog *Leptodactylus* but 5 Na^+ are taken up against 1 NH_4^+. Equation of regression line: $f_{nNH_4^+} = 0.197\ f_{nNa^+} - 0.183$.

frog *Leptodactylus ocellatus* (30). The rates of ammonia excreted are plotted against the rates of sodium uptake (net flux) measured simultaneously in the same animals. No correlation is seen between the two variables in the crayfish. In particular, when the animals are kept in deionized water and no absorption occurs for Na^+, ammonia excretion continues unimpaired or is even augmented. The points on the ordinate axis represent this type of observation (Table 5 in Shaw's paper, 81). Moreover, two further attempts to

observe an obligatory NH_4^+/Na^+ exchange were reported in the same
paper. In one, the animals kept in deionized water were studied.
The rate of ammonia excretion was measured after sudden addition
of Na^+ ions to the external medium and the rate of Na^+ uptake determined simultaneously in the hope to verify Krogh's assertion that
an increase of sodium absorption would be accompanied by additional
ammonia excretion. In point of fact, no such increase was observed,
rather the reverse. In a second type of experiment (given in Table 8
in Shaw's paper), a comparison between the rate of ammonia production and the net uptake of sodium from sodium sulfate solutions was
made. The points on the right hand side of the graph illustrate
these attempts. Shaw concludes that the exchange for ammonium ions
cannot be obligatory, as under certain circumstances the sodium uptake may exceed ammonia production. He does not "rule out the possibility that this type of exchange may constitute the normal mechanism. There is no *a priori* reason for supposing that the exchange
process must necessarily involve only a single ion species." He
suggests that an alternative exchange process may occur through the
same channel, namely a H^+/Na^+ exchange. "Sodium may be exchanged
for either, depending on the relative rates of excretion and the
type of metabolic activity displayed by the animal at the time."

Although no such experiments were attempted on teleosts, we
may safely predict that a similar situation prevails also in fish.
Recently de Vooys (93) observed on the carp transferred from tapwater to deionized water that the ammonia excretion is increased
by about 100%, an increase which is apparent only after about 48 hrs

adaptation. After return to tapwater, the ammonia production decrease again after a delay of 24 - 48 hrs. Obligatory exchange seems therefore to be ruled out. In marine teleosts, the net salt transport is directed outwards and recent experimental evidence suggests that internal Na^+ is exchanged against external K^+ (52). An exchange between endogenous ammonia against external Na^+ is not compatible with the maintenance of mineral balance in the hypersaline medium. Yet marine teleosts are ammonotelic and the rate of ammonia excretion across the gill is of the order of magnitude (up to 20 μmole $\times$ h^{-1} $\times$ 100 g^{-1}) of that reported in freshwater teleosts (86, 33, 34). According to Motais, injection of ammonium salt solution into the seawater adapted eel and flounder is accompanied by an increase of the internal sodium level (60). This observation suggests that the Na^+/NH_4^+ may still be operative in euryhaline fishes kept in sea water. This interesting possibility must be thoroughly substantiated however by additional evidence.

Finally when the biology of the fish is to be considered, some arguments against an obligatory Na^+/NH_4^+ exchange may also become evident. Many observers agree that food intake greatly increases ammonia production and excretion across the gills (26, 93). Yet food intake is also almost certainly a source of additional salt absorption across the gut. The maintenance of the salt balance should logically impose a reduction of the salt uptake across the gills. Obviously changes in diet would impose contradictory demands on the gill exchange mechanisms.

On the right hand side of Fig. 7, it may be seen that a fairly good correlation between Na^+ net uptake and NH_3 excretion occurs in the frog *Leptodactylus*, but in this ureotelic species, the rate of ammonia production falls short of the amount of Na^+ absorbed. Furthermore, injection of ammonia remained without effect in the sodium uptake (31, Garcia Romeu, personal communication). These observations led Garcia Romeu and his colleagues to investigate the alternative possibility: H^+/Na^+ exchange.

Demonstration of a sodium-proton exchange in parallel to HCO_3^-/Cl^- exchange.

Such an exchange was initially suggested by Ussing (91) and in his laboratory Schoffeniels (76, 77) showed that an acidification of the mucosal bathing medium depresses active sodium transport by isolating frog skin. A competition between H^+ and Na^+ for a common carrier is suggested. This problem was reinvestigated later (24, 25) but the studies were made with an external medium containing NaCl. As will be seen below, the HCO_3^-/Cl^- exchange results in an increase of the buffering capacity of the outside milieu and HCO_3^- combines with the protons exchanges against Na^+ to form CO_2 and H_2O. The addition of NH_4^+ ions on the mucosal side increases Na^+ outflux and decreases significantly Na^+ influx, net flux and skin potential. The effect is tentatively ascribed to functional alterations of the transporting epithelial cells (acidification or K^+ loss). Garcia Romeu and his colleagues (31) first demonstrated that Na^+ and Cl^- uptake are independent, especially when the frog *Calyptocephalella gayi* are "preadapted" in salt solutions lacking either Na^+ or Cl^- (see above). This independence is further demonstrated by the possibility of inhibiting specifically Na^+ or Cl^- uptake with inhibitors such as procaine and pentobarbital.

The authors took advantage of this independence in studying the Na^+ and Cl^- transport independently. This was done by placing the frogs (with urinary cloacum catheterized to avoid addition of urine to the bath) in either Na_2SO_4 or choline chloride solutions. Fig. 8

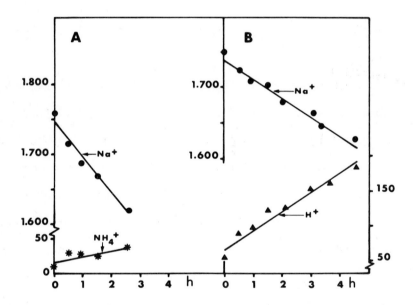

Fig. 8. Comparison between the excretion of cations and the uptake of Na^+ in two different frogs submerged in Na_2SO_4 solutions. A = NH_4^+ excreted against Na^+ taken up $f_{n_{Na^+}}$ = + 10.8 µEq X h^{-1} X 100 g^{-1}, $f_{n_{NH_4^+}}$ = - 1.6 µEq. B = H^+ excreted against Na^+ taken up $f_{n_{Na^+}}$ = + 3.9 µEq, $f_{n_{H^+}}$ = - 3.8 µEq. Ordinate, external concentrations in µEq/liter. Abscissa, time in hr. The fluxes are computed from the slopes

illustrates two typical experiments showing most clearly that it is H^+ excretion and not NH_4^+ production which is to be correlated to Na^+ uptake. As can be seen in Fig. 9, there is an excellent correlation

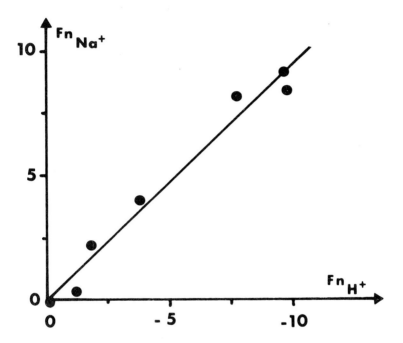

Fig. 9. Regression line between Na^+ uptake and H^+ excreted in seven experiments (31). Ordinate, net uptake flux of Na^+ in µEq X h^{-1} X 100 g^{-1}; Abscissa, net excretion rate of H^+ (same units); Equation, $f_{n_{Na^+}} = 0.99 \, X \, f_{n_{H^+}} - 0.10$.

between proton production and Na^+ net uptake and a one-for-one relationship is the prevailing mechanism. In all the experiments of this kind, a pH downward shift of 0.30 ± 0.04 units per hour is observed

---

(Fig. 8 [p. 136]) of the lines representing the appearance or disappearance of the cations.

in the external medium. This increase of proton activity is small and does not account for the actual proton addition to the bath because of the progressive buffering of the external medium (about + 16 µEq of base per pH unit and per hr). Only total acidity titration with NaOH permitted the evaluation of proton excretion.

Similarly, the endogenous ionic species involved in Cl^- exchange was definitely shown to be HCO_3^- in experiments with frogs sitting in a choline chloride solution. Again an excellent correlation between base excretion and Cl^- uptake is observed but the net flux of base is about 30% higher than anion absorption rate. The pK of the excreted base is identical to that of choline bicarbonate at concentrations equivalent to those of the experimental solution. Finally an increase in buffering capacity (about 76 µEq of acid per unit pH per hr) and of 0.15 ± 0.05 pH units is observed during chloride uptake. The buffering capacity of the excreted base excludes the possibility that it could be hydroxyle ions.

When Na^+ and Cl^- are simultaneously absorbed from NaCl solutions, no pH shift is observed. Sudden addition of H^+ ions to the external bath selectively blocks Na^+ uptake until HCO_3^- excretion has compensated for the pH drop. Such an alkalinization is absent in acidified Na_2SO_4 solutions when no Cl^-/HCO_3^- exchange occurs.

According to Garcia Romeu and his colleagues (31), the Na^+/H^+ exchange just described is suggested to be a general case in the more terrestrial amphibian species, in which a switch from ammonotelism to ureotelism has occurred, with the assumption by the kidney of the function of nitrogenous waste excretion. Replacement of NH_4^+

by H^+ in the ionic exchange mechanism would appear to be linked to ecological changes imposed as a physiological adaptation developed during the transition to life on land. In at least one ammonotelic species, the mosquito larva *Aedes aegypti* (90), a H^+/Na^+ exchange appears to be operative, the NH_4^+/Na^+ mechanism being excluded. Addition of ammonium ions to the outside medium remains without effect on sodium absorption by the anal papillae.

The possibility of a H^+/Na^+ exchange also occurring in the goldfish is suggested by the results of a few preliminary experiments in collaboration with Garcia Romeu. Metabolic acidosis by injection of 0.1 M sulfuric acid solution up to 75 μmole X 100 g^{-1} produced an increase of the sodium influx in one out of two experiments. The sodium net flux remained unchanged however. Addition of H^+ ions (0.1 M sulfuric acid solution) to the outside medium is followed by a partial or total inhibition of Na^+ uptake. In two experiments, a pH decrease of about 2 - 3 units completely inhibited Na^+ transport. Sudden change of pH produced however a significant transient permeability change of the gill epithelium in both directions. It may well be that the observed effects result from unspecific variations in the dissociation of fixed charged in the membrane.

The fact that in neither crayfish nor freshwater teleosts a satisfactory correlation between sodium absorption and ammonia production is obtained may well mean that in these animals too a H^+/Na^+ exchange prevails while ammonia is excreted in molecular form. This hypothesis poses the problem of the membrane permeability to ammonia in molecular or ionic form.

Branchial permeability to unionized or ionized forms of ammonia.

Ammonia exists in solution as ammonium ion (NH_4^+) and ammonia gas (NH_3) according to the following reactions:

$$NH_4^+ \rightleftarrows NH_3 + H^+ \tag{3}$$

$$\frac{NH_3 \times H^+}{NH_4^+} = K \tag{4}$$

$$P_{NH_3} = K \frac{NH_4^+}{H^+} \tag{5}$$

$$pH = pK + \log \frac{NH_3}{NH_4^+} \tag{6}$$

The pK being about 9.3, most of the ammonia in biological fluids (at pH 7.5) or in external media (at pH 6 - 8) is in the ionic form.

Depending on the character of the barrier interposed between ammonia solution, either NH_3 or NH_4^+ or both species may be transported. Non ionic diffusion is a passive process conforming to physico-chemical laws. Because of its high lipid solubility, free ammonia readily traverses most membranes. The net movement of ammonia across such barriers depends therefore on the partial pressure P_{NH_3} on each side, and occurs from higher to lower tension. Free ammonia will pass from the side of a higher pH to the side of a lower pH as is seen in equation (5) and (6) which demonstrates that P_{NH_3} is proportional to pH. Thus a passive transfer of free ammonia may occur even against a concentration gradient of total ammonia (8). Such transfers are furthermore independent of the membrane potential.

Ionic ammonia may be transported either actively or passively. If it can be shown that ammonium crosses a membrane from lower to higher pH, and against an electrochemical gradient, active transport is the prevailing mechanism and cellular energy is required by the membrane for such an operation. This is true unless solvent drag is operative.

In one case -- ammonium absorption by the hamster ileum -- active transport of NH_4^+ has been clearly demonstrated (58). There are several other instances in which a passage of ionic ammonium rather than of free ammonia across membranes has been suggested. In human erythrocytes, NH_4^+ can substitute for K^+ in an inwardly directed active pump (72). NH_4^+ is also able to replace K^+ in activating the sodium-potassium-dependent ATPase (85). This enzyme has also been isolated in the teleostean gill. The observation that it is found in higher quantities in the marine forms has been taken as proof that it is involved in the sodium extrusion mechanism (40, 38, 60, 61), which is probably mediated by a Na^+/K^+ exchange (52, 53). In order to test the possibility of a tied coupling between NH_4^+ extrusion and Na^+ absorption ATPase, Motais (60) compared the affinities of the branchial "freshwater" ATPase for NH_4^+ and K^+. The results were disappointing as the apparent K_m is 3 to 4 times higher for NH_4^+ than for K^+.

NH_4^+ may also substitute for K^+ in inhibiting the sodium-dependent glucose absorption in the hamster small intestine (14). It was also found to be as effective as K^+ as an activator of *in vitro* aldosterone synthesis by mammalian adrenal slices (62).

The mechanism of ammonia excretion in the distal kidney tubule is related to the maintenance of acid-base equilibrium (6, 89, 70). According to the classical model, the distal tubule is responsible for the acidification of the urine by affecting a H^+/Na^+ exchange.

The enzymatic machinery of the tubular cells (especially glutaminase and glutamate dehydrogenase) is activated and augmented amounts of NH_3 presumably in the molecular form pass into the tubular lumen. The protons which originate from the dissociation of H_2CO_3 within the tubular cell and which are secreted simultaneously are captured by NH_3 to form NH_4^+ in the tubular lumen. This "capture by diffusion" mechanism prevents excessive acidification of the urine.

The alternate possibility that in the kidney a NH_4^+/Na^+ exchange is operative is not entirely ruled out (89). Such an exchange is observed in Crocodilia (35). In these animals, the cells of the distal tubules excrete simultaneously ammonium and bicarbonate ions against Na^+ and Cl^- ions which are reabsorbed. The resulting urine being alkaline. NH_4^+ is therefore excreted against a pH gradient.

For the teleostean gill, there are indications that ammonia may cross the epithelium in either molecular or ionic form. Most of the studies concerning toxicity of ammonia in ambient water conclude that the toxicity of any given ammonia solution augments when the external pH is raised. This suggests that it is related to the concentration of free ammonia in the outside medium. Similar observations have been made on the toxicity of ammonia in mammals. Nevertheless when changes in pH are affected by increasing the free CO_2 content of the medium, the toxicity of ammonia augments with decreasing pH (45). According to Warren and Schenker (94) who studied the effects of plasma P_{CO_2} on the toxicity of ammonia in mammals, the permeability of biological membranes to CO_2 and the effect of P_{CO_2} on the pH of the intraepithelial of transepithelial media have to be taken into account. The apparently aberrant results may thus still be explained in terms of permeability to molecular form of NH_3.

Also in accordance with such a passive permeability are preliminary results (95) showing that alkalosis following intraperitoneal injection of $NaHCO_3$ is accompanied by an increase of ammonia excretion while metabolic acidosis produced by injection of HCl is followed by a reduction in the rate of ammonia production. These observations have to be confirmed however, as increased ammonia production in goldfishes during confinement accompanied by a hugh increase of blood lactate concentration has been reported recently (16).

There are indications however that the gill of fish is the site of active transport of ammonia which cannot be explained in terms of simple diffusion of free ammonia. In the report already mentioned above (95), experiments are briefly described concerning the fate of intraperitoneally injected ammonium solutions in the catfish kept in a closed-circuit aquarium. The final concentration of ammonia excreted in the bath was two to three times its concentration in the blood. This occurred even when the blood was more acid than the bath. Excretion two-thirds as efficient as normal against a gradient of 2 $\mu Eq \times liter^{-1}$ has been observed. More recent investigations on the depressing effect of ammonia added to the bath on the ammonia excretion of the rainbow trout suggested however passive permeability of the gill to ammonia (27). The data presented show that there is a direct linear correlation between ambient ammonia and blood ammonia of fish exposed for 24 hrs to varying concentrations of ammonia.

This correlation holds for both total ammonia and free ammonia. Since blood ammonia concentration always exceeded the external water ammonia concentration, the increases in blood ammonia are taken to result from inhibition of NH_3 excretion and this is in harmony with the idea that ammonia is excreted passively. The problem needs to be reinvestigated nevertheless. For instance the ammonia plasma concentrations in the dorsal aorta given in this report are at least 10 times higher than those reported by others (34, 65 - 67, 74).

More work is necessary to bring unequivocal proof in favor of the hypothesis that the gill may be simultaneously the site of passive and active transfer of ammonia. In addition the possibility of biochemical synthesis of ammonia by the epithelial tissue (34) complicates the experimental analysis of the transfer phenomenon and has to be taken into account.

Reappraisal of the alternative model NH_4^+/Na^+ and H^+/Na^+.

If unequivocal demonstration of ammonium transport is obtained, it follows that probably both NH_4^+/Na^+ and H^+/Na^+ exchanges coexist in the gill of aquatic animals together with free-ammonia diffusion. The degree of coupling between the sodium pump and H^+ or NH_4^+ release will be extremely difficult to ascertain.

A first difficulty is to measure in practice H^+ excretion by simple titration of the aquarium water, as released free-ammonia will be trapped in ionized form by excreted respiratory CO_2. In other words simultaneous release of CO_2 and NH_3 is undistinguishable form $NH_4^+ + HCO_3^-$ excretion. In this respect the studies of the carbon dioxide dissociation curves of water breathed and rebreathed by teleost fish are of interest (16). The P_{CO_2} difference between inspired and expired water is small because fish must ventilate strongly to extract poorly-soluble oxygen. This difference is further decreased because in well aerated high carbonate water, the carbonate-bicarbonate system buffers an appreciable amount of CO_2 released and also because for ammonotelic fish titration alkalinity of the external water increases. For Dejours and his colleagues this increase is to be ascribed to ion-exchange by the gill involving NH_4^+ and HCO_3^- production. None of the experiments given by the authors show a net absorption of Na^+ in exchange of NH_4^+. A mutual trapping of NH_3 and CO_2 may well have the same result. It would be of critical interest to follow titration alkalinity in experiments with salt depleted goldfish actively absorbing Na^+ from a Na_2SO_4 solution. In such experimental conditions, providing that the external medium is well aerated to prevent increase of the P_{CO_2}, any downward pH shift despite NH_3 - CO_2 trapping would be strongly indicative of a H^+/Na^+ exchange occurring in the gill. When studying this exchange in frogs, Garcia Romeu and his colleagues (31) were greatly helped because the quantity of ammonia released by the skin was negligible and therefore the titration of the proton release was straightforward.

A second difficulty in checking a stoichiometric relationship between H^+ or NH_4^+ excretion and Na^+ absorption is of a theoretical nature. The problem is to find out which component of the sodium influx is to be correlated to H^+ or NH_4^+ excretion. Logically it should be the *active component* of the influx which should be taken into account. For transporting epithelia however, it is difficult to divide the influx into active, passive and exchange-diffusion components because the Goldman equation relating passive ionic influx to the electrochemical gradient cannot be applied as almost certainly two potential "jumps" occur one at the external and the other at the internal boundaries (53).

In the situation encountered in the study of the frog *in vivo* (31) a one for one relationship between Na^+ *net uptake* and H^+ excretion was observed, while about 2 Cl^- ions (net uptake) were exchanged against 3 HCO_3^- ions. It would have been of interest to have a simultaneous evaluation of the Cl^- or Na^+ influx.

Another difficulty when dealing with inhomogenous membranes such as gill epithelia is that part of the outflux may be across the respiratory lamellae while actually the active influx might be mediated by the mitochondria rich cell situated on the interlamellar region (13). The net flux measured by difference between influx and outflux may not be representative of the active influx component.

In this respect, it may be of interest to recall that in the human red cell a 3 Na^+ for 2 K^+ stoichiometric relationship is observed (32, 72, 85). This Na^+/K^+ exchange concerns only the part of the Na^+ outflux which is K^+_{ext}-dependent and ouabain-sensitive.

For the HCO_3^-/Cl^- exchange a stoichiometric (3 for 2) relationship has been observed in the frog suggesting an obligatory exchange. It may well be that in freshwater teleosts a similar exchange process is in operation. There is further indirect evidence which suggest, for the goldfish, that this exchange is also tightly coupled

(15). When the ionic composition of the external medium is suddenly changed -- transfer from a NaCl to a Na_2SO_4 solution -- a dramatic reduction of the output of respiratory CO_2 is observed within one half hr. In some cases, an absorption of CO_2 by the fish is seen sometimes for as long as 24 hrs. When the animal is replaced in NaCl solution, CO_2 excretion is resumed and considerably increased. The above mentioned variations of the CO_2 transfer across the gill do not reflect variations of the metabolic CO_2 production as O_2 consumption remains essentially steady during the experiment.

According to Dejours, the dependency of the CO_2 transfer upon the ionic composition of the water is best explained in terms of an obligatory Cl^-/HCO_3^- exchange: HCO_3^- entry against Cl^- loss following the suppression of external Cl^-, or massive HCO_3^- exit upon return to NaCl solution and compensatory Cl^- uptake induced by either internal Cl^- depletion or elevated plasma $[HCO_3^-]$ concentration. This interpretation suggests that most of CO_2 release is not in molecular but in the ionized form.

This experiment also demonstrates the role of the gill in the maintenance of the acid-base balance. A similar conclusion was drawn for the skin by Garcia Romeu and his colleagues (31) from their experiments on H^+/Na^+ and Cl^-/HCO_3^- exchange in the frog.

CONCLUSION

Recent experiments confirm the possibility that in freshwater animals ammonia excretion (ammonotelism) may be related to the maintenance of the acid-base balance and to cation conservation as suggested by Homer Smith many years ago. The evidence is however mainly indirect: addition of ammonia to the outside milieu is followed by an inhibition of sodium uptake and injection of ammonia is followed by an increase of sodium uptake. In parallel a HCO_3^-/Cl^- exchange is operative. Direct demonstration of a stoichiometric relationship between NH_3 output and Na^+ uptake has not been obtained. In some cases ammonia excretion continues in the absence of Na^+ uptake and in others the rates Na^+ uptake is greater than NH_3 excretion. The possibility that the gill is permeable to both molecular and ionized forms of ammonia is discussed in the light of experimental evidence.

In ureotelic amphibia the rate of Na^+ uptake greatly exceeds ammonia excretion by the skin. In this case a H^+/Na^+ exchange is operative. It may well also occur in ammonotelic animals. The difficulties in choosing between H^+/Na^+ and NH_4^+/Na^+ exchanges in parallel with NH_3 (molecular) excretion are pointed out.

SUMMARY

Recent observations, obtained mainly on fish and crustacea, suggest that in freshwater animals excretion of ammonia by the gill (ammonotelism) is a physiological function linked to the maintenance of acid-base balance and to the cation (sodium) absorption mechanism as suggested by H. Smith and A. Krogh many years ago.

Experimental evidence in favor of a Na^+/NH_4^+ exchange is however mainly indirect. Addition of ammonium salts to the external medium results in an inhibition of sodium absorption, while injection of ammonium salts into the internal medium produces an increase of the sodium uptake, in accordance to the suggested Na^+/NH_4^+ exchange model. In parallel, a HCO_3^-/Cl^- exchange is observed.

A direct demonstration of such tightly linked coupled exchanges has not been obtained. No stoichiometric relationship between the rates of ammonium excretion and sodium absorption is observed. For instance, excretion of ammonia continues in animals kept in deionized water, in the absence of cation absorption. In some cases, the rate of ammonia excretion is much higher than that of sodium absorption, while in others the reverse situation occurs.

Present evidence suggests that the gill may be permeable to both ionized and molecular forms of ammonia and that both active and passive transport of this substance coexist in the gill. Furthermore, the possibility of a H^+/Na^+ exchange is also suggested. Such an exchange has been shown to occur in ureotelic amphibians kept in freshwater, in which the rate of sodium absorption by the skin is always much greater than that of ammonia excretion. The difficulty to distinguish between H^+/Na^+ in parallel to NH_3 (molecular) excretion and NH_4^+/Na^+ exchange is emphasized.

ACKNOWLEDGEMENTS

The author wishes to thank Dr. Garcia Romeu for very stimulating discussions and Mr. Robert Langford for correcting the manuscript.

REFERENCES

1. Aceves, J., D. Erlij and C. Edwards. *Biochim. Biophys. Acta 150*: 744, 1968.

2. Alvarado, R.H. and T.H. Dietz. *Comp. Biochem. Physiol. 33*:93, 1970.

3. Alvarado, R.H. and A. Moody. *Amer. J. Physiol. 218*:1510, 1970.

4. Baldwin, E. *An Introduction to Comparative Biochemistry.* Cambridge University Press, 1949.

5. Balinsky, J.B. and E. Baldwin. *J. Exp. Biol. 38*:695, 1961.

6. Berlinger, R.W. *Circulation 21*:892, 1960.

7. Bourguet, J., B. Lahlou and J. Maetz. *Gen. Comp. Endocrinol. 4*:563, 1964.

8. Bromberg, P.A., E.D. Robin and C.E. Forkner, Jr. *J. Clin. Invest. 39*:332, 1960.

9. Brown, A.C. *J. Cell. Comp. Physiol. 60*:263, 1962.

10. Bryan, G.W. *J. Exp. Biol. 37*:83, 1960.

11. Chaisemartin, C. *C.R. Soc. Biol. 160*:1305, 1966.

12. Chaplin, A.E., A.K. Huggins and K.A. Munday. *Comp. Biochem. Physiol. 16*:49, 1965.

13. Conte, F.P. In: *Fish Physiology, Vol. I*, edited by W.S. Hoar and D.J. Randall. New York: Academic Press, 1969. p. 241.

14. Crane, R.K. *Fed. Proc. 24*:1000, 1965.

15. Dejours, P. *J. Physiol. (London) 202*:113 P, 1969.

16. Dejours, P., J. Armand and G. Verriest. *Resp. Physiol. 5*:23, 1968.

17. Delaunay, H. *C.R. Soc. Biol. 101*:371, 1929.

18. Delaunay, H. *Biol. Rev. 6*:265, 1931.

19. Dietz, T.H., L.B. Kirschner and D. Porter. *J. Exp. Biol. 46*:85, 1967.

20. Downing, K.M. and J.C. Merkens. *Annls. Appl. Biol. 43*:243, 1955.
21. Fanelli, G.M. and L. Goldstein. *Comp. Biochem. Physiol. 13*:193, 1964.
22. Forster, R.P. and L. Goldstein. In: *Fish Physiology, Vol. I,* edited by W.S. Hoar and D.J. Randall. New York: Academic Press, 1969. p. 313.
23. Florkin, H. and G. Frappez. *Arch. Int. Physiol. 50*:197, 1940.
24. Friedman, R.T., R.M. Aiyawar, W.D. Hughes and E.G. Huf. *Comp. Biochem. Physiol. 23*:847, 1967.
25. Friedman, R.T., R.N. Laprade, R.M. Aiyawar and E.F. Huf. *Amer. J. Physiol. 212*:962, 1967.
26. Fromm, P.O. *Comp. Biochem. Physiol. 10*:121, 1963.
27. Fromm, P.O. and J.R. Gillette. *Comp. Biochem. Physiol. 26*:887, 1968.
28. Garcia Romeu, F. and J. Maetz. *J. Gen. Physiol. 47*:1195, 1964.
29. Garcia Romeu, F. and R. Motais. *Comp. Biochem. Physiol. 17*:1210, 1966.
30. Garcia Romeu, F. and A. Salibian. *Life Sci. 7*:465, 1968.
31. Garcia Romeu, F., A. Salibian and S. Pezzani-Hernandez. *J. Gen. Physiol. 53*:816, 1969.
32. Glynn, I.M. *British Med. Bull. 24*:165, 1968.
33. Goldstein, L. and R.P. Forster. *Amer. J. Physiol. 200*:1116, 1961.
34. Goldstein, L., R.P. Forster and G.M. Fanelli. *Comp. Biochem. Physiol. 12*:489, 1964.
35. Hernandez, T. and R.A. Coulson. *Science 119*:291, 1954.
36. House, C.R. *J. Exp. Biol. 40*:87, 1963.

37. Huggins, A.K., G. Skutsch and E. Baldwin. *Comp. Biochem. Physiol.* 28:587, 1969.

38. Jampol, L.M. and F.H. Epstein. *Amer. J. Physiol.* 218:607, 1970.

39. Jorgensen, C.B., H. Levi and K. Zerahn. *Acta Physiol. Scand.* 30:178, 1954.

40. Kamiya, M. and S. Utida. *Comp. Biochem. Physiol.* 31:671, 1969.

41. Kirschner, L.B. *J. Cell. Comp. Physiol.* 45:61, 1955.

42. Krogh, A. *Osmotic Regulation in Aquatic Animals.* Cambridge University Press, 1939.

43. Leiner, M. *Z. vergl. Physiol.* 26:416, 1938.

44. Leloup, J. *J. Physiol. (Paris)* 58:560, 1966.

45. Lloyd, R. and D.W.M. Herbert. *Annls. Appl. Biol.* 48:399, 1960.

46. Lockwood, A.P.M. *Aspects of the Physiology of Crustacea.* Olivers and Boyd, Edinburgh and London, 1968.

47. MacBean, R.L., M.J. Neppel and L. Goldstein. *Comp. Biochem. Physiol.* 18:909, 1966.

48. MacFarlane, N. and J. Maetz, in preparation.

49. Maetz, J. *Bull. Inst. Oceanog. Monaco* 43:1, 1946.

50. Maetz, J. *Bull. Biol. France and Belgique Suppl.* 40:1, 1956.

51. Maetz, J. *J. Physiol. (Paris)* 48:1085, 1956.

52. Maetz, J. *Science* 166:613, 1969.

53. Maetz, J. *Bull. Inf. Sci. Techn. C.E.A.* 145:3, 1970.

54. Maetz, J. *Mem. Soc. Endocrinol.* 18:3, 1970.

55. Maetz, J. and G. Campanini. *J. Physiol. (Paris)* 58:248, 1966.

56. Maetz, J. and F. Garcia Romeu. *J. Gen. Physiol.* 47:1209, 1964.

57. Maetz, J. and G. Zwingelstein, unpublished.

58. Mossberg, S. *Amer. J. Physiol. 213*:1327, 1967.
59. Motais, R. *Annls. Inst. Oceanog. Monaco 45*:1, 1967.
60. Motais, R. *Bull. Inf. Sci. Techn. C.E.A. 146*:3, 1970.
61. Motais, R. *Comp. Biochem. Physiol. 34*:497, 1970.
62. Müller, J. *Nature (London) 206*:92, 1965.
63. Needham, A.E. *Biol. Rev. 13*:224, 1938.
64. Parry, G. In: *The Physiology of Crustacea, Vol. I*, edited by T.H. Waterman. New York: Academic Press, 1960. p. 341.
65. Pequin, L. *C.R. Acad. Sci. 255*:1795, 1962.
66. Pequin, L. *Arch. Sci. Physiol. 21*:193, 1967.
67. Pequin, L. and A. Serfaty. *Comp. Biochem. Physiol. 10*:315, 1963.
68. Pequin, L. and A. Serfaty. *Arch. Sci. Physiol. 22*:449, 1968.
69. Pickford, G.E. and F.B. Grant. *Science 155*:568, 1967.
70. Pitts, R.F. *Amer. J. Med. 36*:720, 1964.
71. Pora, E.A. and A. Precup. *J. Physiol. (Paris) 50*:459, 1958.
72. Post, R.L. and P.C. Jolly. *Biochim. Biophys. Acta 25*:118, 1957.
73. Potts, W. *Amer. Rev. Physiol. 30*:73, 1968.
74. Robertson, J.D. *J. Exp. Biol. 31*:424, 1954.
75. Salibian, A., S. Pezzani-Hernandez and F. Garcia Romeu. *Comp. Biochem. Physiol. 25*:311, 1968.
76. Schoffeniels, E. *Arch. Int. Physiol. Biochem. 63*:513, 1955.
77. Schoffeniels, E. *Arch. Int. Physiol. Biochem. 64*:571, 1956.
78. Schoffeniels, E. *Arch. Int. Physiol. Biochem. 76*:319, 1968.
79. Sharma, M.L. *Comp. Biochem. Physiol. 19*:681, 1966.
80. Shaw, J. *J. Exp. Biol. 36*:126, 1959.
81. Shaw, J. *J. Exp. Biol. 37*:534, 1960.

82. Shaw, J. *J. Exp. Biol. 37*:548, 1960.
83. Shaw, J. *J. Exp. Biol. 37*:557, 1960.
84. Shaw, J. *Symp. Soc. Exp. Biol. 18*:237, 1964.
85. Skou, J.C. *Physiol. Rev. 45*:596, 1965.
86. Smith, H.W. *J. Biol. Chem. 81*:727, 1929.
87. Smith, H.W. *Quarterly Rev. Biol. 7*:1, 1932.
88. Smith, H.W. *From Fish to Physiology*. Oxford University Press, New York, 1956.
89. Smith, H.W. *Principles of Renal Physiology*. Oxford University Press, New York, 1956.
90. Stobbart, R.H. *J. Exp. Biol. 47*:35, 1967.
91. Ussing, H.H. *Acta Physiol. Scand. 17*:1, 1949.
92. Ussing, H.H. *Handbuch Exptl. Pharmakol. 13*:1, 1962.
93. Vooys, G.G.N., de. *Arch. Int. Physiol. Biochem. 76*:268, 1968.
94. Warren, S.K. and S. Schenker. *Amer. J. Physiol. 203*:903, 1962.
95. Wolbach, R.A., H.O. Heinmann and A.P. Fishman. *Bull. Mt. Desert Isl. Biol. Lab. 4*:56, 1959.
96. Wuhrmann, K. von and H. Woker. *Schweiz. Z. Hydrol. 11*:210, 1968.

AMINO ACID ACCUMULATION AND
ASSIMILATION IN MARINE ORGANISMS[1]

Grover C. Stephens

*Department of Developmental and
Cell Biology,
University of California,
Irvine, California 92664*

Free amino acids (FAA) dissolved in sea water can and do meet nutritional requirements of marine organisms. Some requirements can be met in full by naturally occurring FAA, in other cases FAA provides a substantial supplement to the total intake of organic matter. Not all organisms are adapted to take advantage of this potential food source. However, the capacity for amino acid accumulation of FAA may be a phenomenon of general importance. The present paper attempts to establish an important role for FAA in selected cases and discusses some characteristics of uptake and utilization of dissolved amino acids.

A brief discussion of marine environments is necessary. Marine environments are diverse although emphasis is usually placed on uniformity with respect to temperature, salinity, oxygen content and

---

[1]Unpublished observations reported in this paper were supported by Research Grant GM-12889 from the USPHS and by Research Grant GB-17263 from the NSF.

other environmental factors in the open ocean. However, in inshore areas a great diversity of environments is available to marine organisms. Oxygen content may drop from saturation to zero a few millimeters below the sediment surface in a sand-mud bottom; redox potential drops by half a volt. Intertidal and estuarine habitats also present dramatic contrasts. Excess oxygen during the day and anaerobic conditions at night, alternate freezing and thawing with the movement of the tides, and sweeping salinity changes in estuaries are familiar examples. Even the apparently homogeneous water mass of the open ocean exhibits heteogeneity. It is stratified and has small domains which differ in composition from surrounding water and move as differentiated units. Such differences are invoked to account for the local character of plankton distribution.

The diversity of marine environments includes differences in the dissolved organic material (DOM) in sea water. The mass of unorganized organic matter usually exceeds biomass in a community. This is also true of marine habitats and communities. Total organic carbon amounts to roughly 1 mg/liter in surface water of the ocean and decreases to perhaps half that value in deeper waters (6, 8, 9, 19, 25). Particulate matter is an order of magnitude less than DOM at the surface and its relative contribution decreases in deeper waters. Carbon-nitrogen ratios of DOM vary but are low enough to justify the conclusion that much of the organic material contains nitrogen. Free amino acids in ocean water comprise roughly 5% of the total DOM (6). This is roughly equivalent to a total concentration of 5×10^{-7} moles/liter though somewhat higher concentrations may occur seasonally (3). There is considerable diversity in reports of FAA from surface waters due to the analytical problems presented in their determination (see Bohling [3] for a recent discussion). However, there is evidence for regional and seasonal differences.

Differences in DOM between open water and sediment habitats are striking. According to Degens (6), organic matter comprises 0.1 to 3% of the weight of marine sediments. The concentration of nitrogen-containing compounds is high near the surface and decreases with depth. Thus the total organic material in sediment exceeds that of surface waters by three to four orders of magnitude. FAA extracted with water or ammonium acetate from sediments ranges from 10 mg to 1 gm per kilogram or 1 to 3% of the total. This is roughly 10^{-2} to 10^{-4} moles/liter. A portion of this FAA is probably sequestered in free amino acid pools of sedimentary organisms or absorbed on organic or inorganic particles. However, a portion is in solution in the interstitial water of the sediment. Stephens (39) reports concentrations of 6×10^{-5} to 1.1×10^{-4} for the interstitial water of sand-mud sediment. These figures probably represent a few percent of the extractable FAA in such sediments. I am not aware of other published figures for FAA in interstitial water.

The sources of organic material in marine environments are not well understood. Little is known about the character and rates of the dynamic processes which produce steady state concentrations of DOM in open water and sediments. Nonetheless, there are very large differences in total organic content and in concentration of FAA in different marine habitats. These must be taken into account when assessing the potential significance of amino acid assimilation and accumulation for a particular organism in a specific habitat.

AMINO ACID INFLUX

Stephens and Schinske (43) reported net movement of amino acids from solution in sea water into 35 genera of marine invertebrates, representing 11 phyla. Disappearance of glycine (and other amino acids) from sea water was followed colorimetrically. Concentrations of 2×10^{-3} to 2×10^{-5} moles/liter were employed. Technical limitations prevented use of lower ambient concentrations. However, this directly demonstrates net influx at concentrations below 20 µmoles per liter of amino acid.

Subsequently, a series of papers has appeared (1, 2, 4, 15, 21, 26, 27, 28, 29, 30, 32, 33, 34, 38, 39, 40, 41, 42, 43, 44, 45, 48, 49) reporting the characteristics of amino acid influx into invertebrates and algae. In this work the disappearance of ^{14}C-labelled amino acid from the medium and its appearance in various chemical fractions of the organism was measured. The information concerning influx from these publications can be summarized as follows:

1. Labelled amino acid disappears rapidly from solution and can be recovered quantitatively from the organism. Concentrations as low as 5×10^{-9} moles/liter were used.
2. Rapid entry of amino acids characterized all soft-bodied animals studied, though rates vary. Exceptions are arthropods (1, 2) and vertebrates other than hagfishes (15).
3. Amino acids enter across the body surface generally (34, 38, 39) or specialized regions such as the ctenidia of

molluscs. Little enters by way of the gut.

4. In most cases, the system which mediates influx seems to involve a carrier or transfer site for amino acid transport from the medium to the interior of the organism. This is based on the following considerations:

   a. The systems mediating influx in most cases are saturable and are well described by Michaelis-Menten kinetics.

   b. In some cases, amino acid intake is mediated by more than one system, each of which is specific for a group of amino acids and relatively insensitive to the presence of others.

5. In the saturable accumulation systems which have been studied, the concentrations at which entry occurs at half the maximum rate (K_m or K_s) range from 5×10^{-5} to 2×10^{-3} moles/liter for marine invertebrates. Maximum velocity of uptake is 10^{-7} to 10^{-5} moles/g/hour. K_m's are lower and V_{max} higher for phytoplankters studied.

6. Influx continues unabated for long periods (up to several days) in both algae and marine invertebrates (27, 42).

ASSIMILATION

Accumulated amino acids participate in synthetic and respiratory metabolism of the organism (*cf.* references cited and Fig. 1). Interconversion among the amino acids in the FAA pool of the organism can also be demonstrated. Several investigators have also reported the incorporation of ^{14}C supplied as labelled amino acid into water and

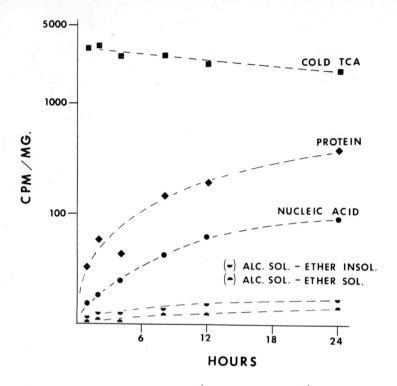

Fig. 1. Incorporation of ^{14}C supplied as ^{14}C-glycine to the brittle star, *Ophiactis*, for one hour and then transferred to sea water. Fractions are: cold 5% trichloroacetic acid, 75% aqueous ethanol, ether, boiling 5% TCA (nucleic acid fraction). The residue is termed residual protein.

alcohol-insoluble fractions of marine organisms using autoradiography (11, 12, 13, 22, 36). This has been verified for several other forms in our laboratory (unpublished).

EFFLUX OF AMINO ACID

There are many reports of amino acid efflux from marine organisms

Examples are: Fogg (14), algae; Hammen et al. (16) and Potts (31), molluscs; Hellebust (17), phytoplankters; Stephens (42) and Wong (48), annelids; Johannes et al. (20), flatworms; Webb and Johannes (46), zooplankton. This is a partial list and could easily be extended by inclusion of earlier reports and reports of amino acid efflux from vertebrates, higher plants, and microorganisms. Unfortunately some of the older reports are based on procedures which did not exclude the possibility of microbial contributions of nitrogen excretion (cf. 7, 37).

Johannes et al. (20) measured leakage of FAA by determining non-volatile ^{14}C released from the flatworm, Bdelloura, and also by direct chemical determination (ninhydrin reaction) of FAA liberated from large groups of animals. The highest rate they report is 130 μg/g/day. Leakage rates estimated by Stephens (42) and Wong (48) for the annelid, Stauronereis, were obtained by measurements of non-volatile ^{14}C released to the medium. Leakage for the five amino acids studied is approximately 400 μg/g/day. Johannes and co-workers correctly note that this approach depends on the assumption that the FAA pool is uniformly labelled. Ferguson (11, 13) presents evidence that assimilation of ^{14}C-amino acids is restricted to superficial tissues in starfishes and Pequignat and Pujol (30) report a rather slow equilibration in sea urchins. Equilibration may be more rapid in annelids (unpublished observations) but the assumption of uniform labelling is questionable. Therefore the efflux rates reported by Johannes et al. have more weight and probably are more reliable.

Despite reservations about leakage measurements based on the appearance of non-volatile ^{14}C in the medium, they have the advantage that they can be made in the presence of high concentrations of unlabelled external amino acid. A chemical determination could not detect the small changes due to efflux at these concentrations. Stephens (42), Wong (48) and Wong and Stephens (49) studied changes in efflux rates of amino acids and amino acid analogues in the presence of various external concentrations of FAA in *Stauronereis*. Rates of efflux did not appear to increase significantly in the presence of 10^{-4} moles/liter of the amino acid used (42). Subsequent investigation supports the generalization that ambient amino acid concentrations have very little influence on efflux rates. Table 1 reports data on efflux rates of the amino acid analogue 1-amino-cyclopentane-1-carboxylic acid (cycloleucine) as a function of internal pool size and concentration of amino acids in the medium. These data were obtained using a continuous flow system which passed sea water through a chamber containing worms previously exposed to ^{14}C-labelled cycloleucine. In general, the influence of external concentration is small compared to the influence of internal cycloleucine levels. Efflux is increased when the worms are exposed to very high concentrations (10^{-2} moles/liter) of arginine or alanine. Alanine is an effective competitive inhibitor of uptake of cycloleucine while arginine is not. The fact that both amino acids are effective at very high concentration is promoting efflux suggests that this action is not a specific one; *i.e.*, it is not a carrier mediated exchange diffusion.

Table 1. *Leakage rates of cycloleucine into solution of amino acid in sea water*

Ambient solution (moles/l)	Cycloleucine concentration in *Stauronereis*	Leakage rate ± S.D. (moles/g/hr)
Sea water	0.80×10^{-3}	$3.66 \times 10^{-10} \pm 0.75$
10^{-4} cycloleucine	0.80×10^{-3}	$6.95 \times 10^{-10} \pm 1.24$
Sea water	1.08×10^{-3}	$7.72 \times 10^{-10} \pm 2.45$
10^{-4} alanine	1.08×10^{-3}	$6.88 \times 10^{-10} \pm 1.36$
Sea water	1.08×10^{-3}	$1.69 \times 10^{-10} \pm 0.46$
10^{-4} arginine	1.08×10^{-3}	$2.03 \times 10^{-10} \pm 0.58$
Sea water	4.58×10^{-3}	$10.10 \times 10^{-10} \pm 3.80$
10^{-4} alanine	4.58×10^{-3}	$15.10 \times 10^{-10} \pm 3.00$
10^{-2} alanine	4.58×10^{-3}	$66.50 \times 10^{-10} \pm 14.4$
Sea water	3.86×10^{-3}	$11.10 \times 10^{-10} \pm 0.48$
10^{-4} arginine	3.86×10^{-3}	$16.70 \times 10^{-10} \pm 9.80$
10^{-2} arginine	3.86×10^{-3}	$75.40 \times 10^{-10} \pm 53.1$
Sea water	8.76×10^{-3}	$41.90 \times 10^{-10} \pm 23.4$
10^{-4} cycloleucine	8.76×10^{-3}	$46.60 \times 10^{-10} \pm 14.6$
Sea water	9.83×10^{-3}	$113.0 \times 10^{-10} \pm 41.0$
10^{-4} arginine	9.83×10^{-3}	$112.0 \times 10^{-10} \pm 23.0$

AMINO ACID AS A NUTRITIONAL SUPPLEMENT IN SEDIMENTARY INVERTEBRATES

Stephens (39) estimated the relative importance of amino acid uptake by comparing it to the metabolic requirements of the annelid, *Clymenella torquata*. Oxygen consumption in this worm is equivalent to the complete oxidation of approximately 90 µg amino acid per g per hr. FAA in the habitat was measured and the rate of accumulation for each amino acid calculated from laboratory data. The worms could acquire 135 µg amino acid per g per hr from the amino acid mixture found in the interstitial water where they were collected. This calculation does not defend the position that amino acid uptake is the only source of reduced carbon. Rather it is an estimate of whether the magnitude of input from the FAA is large or small compared to total requirements. There are several possible sources of error. The most serious is that the worms are facultative anaerobes. If these animals derive some of their energy by anaerobic pathways, their requirement for reduced carbon will increase due to a decrease in efficiency of utilization. Also, if they spend a substantial portion of time under anaerobic conditions, comparison of inputs with oxygen consumption is not reasonable. Most workers believe these worms irrigate their tubes and are aerobes despite the lack of oxygen in the sediments in which they live (*cf*. Mangum [23]). However, this must be borne in mind in generalizing about FAA as a source of reduced carbon for other sediment dwelling animals.

Other sources of error appear to be small. Interactions between amino acids can influence rates of influx. Also, it is possible that some other constituent in the interstitial water might inhibit uptake

However, measurements of uptake of trace amounts of labelled amino acid from interstitial water samples were not significantly different from rates observed in various artificial sea water mixtures or in filtered surface water. Short periods of oxygen deficiency do not decrease influx; uptake was found to continue unabated when the worms were kept in nitrogen saturated sea water.

Another source of error is the efflux of amino acid. The question is whether leakage of FAA exceeds intake or is a large fraction of intake for a particular organism under normal conditions. Rates of influx appear to be much higher than efflux at FAA levels found in natural habitats. *Clymenella* can apparently obtain over 3 mg amino acid per day. Loss has not been measured for this animal but efflux from *Stauronereis* is an order of magnitude less and is not influenced by external amino acid concentrations in the normal range. Influx rates are comparable for the two worms. The ambient concentration of several amino acids at which efflux equals influx in *Stauronereis* is presented in Table 2. When an ambient concentration exceeds the value listed, the animal shows a net uptake of amino acid. The same table presents measured concentrations of those amino acids in interstitial water and the ratio of the existing level in the environment to that at which influx and efflux are equal.

However, Johannes et al. (20) found that efflux exceeds in *Bdelloura* at an ambient amino acid concentration of 60 µg/liter (about 5×10^{-7} moles/liter). They further suggested that exchange diffusion might contribute substantially to apparent influx rates

Table 2. *Concentration of amino acid at which influx equals efflux in* Stauronereis *compared with concentrations in interstitial water of sand-mud sediments*

Amino acid	Concentration (influx = efflux) moles/liter	Concentration in habitat*	Ratio
Alanine	5.0×10^{-8}	1.6×10^{-5}	313
Aspartate	6.5×10^{-8}	8.7×10^{-6}	134
Glutamate	1.8×10^{-7}	2.5×10^{-5}	139
Glycine	7.5×10^{-7}	9.6×10^{-6}	13
Phenylalanine	1.0×10^{-7}	2.5×10^{-6}	25
Valine	1.8×10^{-8}	--	--

*Data from Stephens (39).

as determined by uptake of ^{14}C-labelled compounds. The FAA pool of marine invertebrates is very large compared to ambient concentrations Hence the specific activity of an amino acid would be high in the medium and very low in the animal which tends to obscure exchange diffusion. However, large increases in the concentration of ^{12}C amino acid in the medium does not modify the rate of loss substantially. Also, leakage continues into solutions which contain no amino acid. Thus exchange diffusion is probably not an important aspect of the process.

If efflux is independent of external concentration and influx is linearly dependent upon it (at concentrations well below the K_m), it is possible to choose a concentration such that efflux exceeds influx. Thus there is no reason the conclusion of Johannes and coworkers that efflux exceeds influx in *Bdelloura* in an ambient solution of 5×10^{-7} moles/liter. The ratio of efflux to influx would presumably be still greater at a lower ambient concentration and influx would exceed efflux at higher concentrations. This assumes *Bdelloura* has a constant efflux rate like the worms we have studied. Johannes et al. (20) present no direct evidence to the contrary. What is an appropriate external amino acid concentration to consider in the case of *Bdelloura*? The animal is found on gill leaflets of the horseshoe crab, *Limulus*. *Limulus* feeds on sedimentary invertebrates. In addition, the horseshoe crab may lose amino acids as do other arthropods studied by Webb and Johannes (46). The gill is a likely pathway for such loss. Thus it is difficult to decide what the amino acid concentration may be in the microhabitat of *Bdelloura*. Johannes and co-workers selected an external concentration of 60 µg per liter based on measurements of amino acid in surface waters (5, 47). Though this may prove to be correct, it will be difficult to establish for the ectocommensal animal they have studied. Failing direct measurement of the FAA available to an animal living on the gill of *Limulus*, it does not seem *prima facie* reasonable to equate it to the available FAA in surface waters of the York River Estuary (47).

More data are needed to establish the FAA content of interstitial water in sediments. We have recently undertaken to investigate this matter. Water samples are collected by placing a dialysis bag containing sterile sea water in the habitat to be examined. After the diffusible components of the system have equilibrated, it is retrieved, checked for continued sterility, and the FAA concentration determined. Total ninhydrin positive material (with and without precipitation with cold 5% TCA) has already been determined for 19 such samples. Values range from 46 to 220 μmole FAA/liter. Three determinations for water in the algal mat where *Stauronereis* is found average 250 μmole/liter. These figures are in reasonable agreement with the earlier report of 60 to 110 μmole/liter (38). A quantitative inventory is not yet available but TLC shows the presence of 15 amino acids in these recent samples.

We can summarize the evidence supporting a substantial influx of FAA into sediment-dwelling invertebrates. Influx from solutions of 20 μmole/liter has been demonstrated by following the disappearance of ninhydrin positive material for the peanut worm, *Golfingia*, the clam, *Spissula*, and the sea cucumber, *Thyone* (43). Analyses of total FAA in the habitat of such animals range from 5×10^{-5} to 2×10^{-4} moles/liter. At these concentrations, influx exceeds efflux by an order of magnitude. It would be very desirable to have data for influx, efflux, and naturally occurring levels of FAA in the microhabitat for the same organism. We are in the process of obtaining this information for the annelid, *Stauronereis rudolphi*, and will report elsewhere. Preliminary data support the position that this animal has a substantial net intake of amino acid from the environment.

AMINO ACID AS A NUTRITIONAL SUPPLEMENT IN PHYTOPLANKTON

The FAA in the habitat of a phytoplankter is more firmly

established; a reasonable range to consider is from 5×10^{-7} to 3×10^{-6} moles FAA/liter (3, 5, 6, 47). Several investigators have reported uptake of ^{14}C-labelled amino acid by microscopic algae (18, 26, 27, 28, 35). The measurements of North and Stephens provide sufficient information to estimate the potential significance of amino acid uptake in *Platymonas*, a green flagellate found in phytoplankton in inshore locations and common in tide pools.

Two methods were used to estimate the nutritional contribution made by FAA uptake. The first assumes that FAA contributes to the nitrogen requirements of the cells. *Platymonas* was first shown to grow in fixed volume culture on glycine as the sole nitrogen source (supplied at 2×10^{-3} moles/liter). Nitrogen content, generation time, and rate of uptake of ^{14}C-glycine supplied at 10^{-6} moles/liter were then determined. Nitrogen content and generation time establish the minimum rate at which nitrogen must be supplied to the cells to support growth. This can be compared to the measured rate of ^{14}C-glycine uptake using the following relation (*cf.* Fencl [10]).

$$dU/dt = kN_0 e^{rt}$$

where

U = uptake in moles

N_0 = cell number at time 0

r = specific growth rate (ln 2/doubling time)

k = uptake rate

t = time.

Integrating and supplying the measured values for nitrogen content, doubling time, and uptake rate, the percentage of nitrogen required for cell growth which could be supplied by uptake from 10^{-6} moles per liter can be calculated. The cells from fixed volume culture contained about 7% nitrogen (dry weight), had a generation time of 22 hours, and acquired 3.4 nmole of glycine per min per 5×10^5 cells from solution. This represents approximately 10% of the nitrogen required to support the observed growth.

Additional experiments were done with a continuous flow culture apparatus. Nutrient levels in the culture were modified by changing the rate at which medium was supplied (dilution rate). As dilution rate decreases, cell nitrogen content decreases, doubling time increases, and the rate of glycine uptake increases. The net effect of these changes is to increase the contribution of amino acid uptake to nitrogen requirements. At slower growth rates (low dilution rate) 260% of the nitrogen required for doubling can be supplied by glycine uptake from a solution of 5×10^{-7} moles/liter. Table 3

Table 3. *The contribution of uptake of glycine from 5×10^{-7} moles/liter to the nitrogen required for growth of* Platymonas *in continuous culture at different dilution rates*

Generation time (hours)	Uptake rate (µg N/hr/10^8 cells)	Cell N (µg N/10^8 cells)	Nitrogen contribution (see text)
51	0.7	840	6%
--	1.3	700	-
--	5.3	670	-
164	7.2	660	260%

illustrates these relations (recalculated from data presented by North and Stephens [27]).

Low nutrient levels, longer generation times, and lower nitrogen content are more closely comparable to natural conditions. The observation that glycine uptake by *Platymonas* was greatly stimulated by restricting nutrients led to investigation of other phytoplankters. Uptake of amino acid was studied in several organisms (Table 4) grown

Table 4. *Amino acid accumulation by phytoplankters grown under high (2×10^{-3} g atom N/liter) and low (2×10^{-4} g atom N/liter) nitrogen culture conditions*[a]

Organism	Amino acid	% ^{14}C-amino acid removed after 30 min	
		high N	low N
Chlorella sp.	glycine[b]	< 1	< 1
Chlorella sp.	arginine	37	82
Melosira sp.	arginine	15	28
Nitzschia closterium	glycine	< 1	5
Nitzschia ovalis	glycine	1	24
Thalassiosira flaviatilis	glycine[b]	< 1	< 1
Thalassiosira fluviatilis	arginine	4	24

[a]Data from North, Stephens and North (28).

[b]Glycine was not an effective nitrogen source in supporting growth of these organisms.

in batch culture under high nitrogen (2×10^{-3} g atom N/liter) and low nitrogen (2×10^{-4} g atom N/liter) conditions. Cells were harvested in the late log phase. Low nitrogen conditions stimulated amino acid uptake in several phytoplankters. Increased rates of uptake have been demonstrated for several different amino acids. V_{max} is significantly increased under low nitrogen conditions while K_m remains unchanged. North and Stephens (unpublished) have shown that other nitrogen sources (nitrate and ammonia) do not interfere with uptake and assimilation of amino acids.

Leakage of organic material from phytoplankters has been studied by Hellebust (17). The levels reported for *Platymonas* (*Tetraselmis*) are very low; 1% of photoassimilated carbon excreted during a 48-hour period. Even if all the organic carbon lost was nitrogen containing (which is not likely), this would represent less than a 5% increase in the nitrogen requirements used for our calculations. Thus it appears that FAA in the environment may be a significant source of nitrogen for *Platymonas* and other phytoplankters, even at the low concentrations characteristic of surface waters.

The amino acid molecule may also provide a source of reduced carbon to the phytoplankters. Dark respiration of *Platymonas* ranges from 1.3 to 3.5 µmole O_2/hr/10^8 cells. Populations from continuous culture can acquire 0.5 µmole glycine/hr/10^8 cells from an external concentration of 5×10^{-7} moles/liter. Complete oxidation of this amount of glycine requires about 0.9 µmoles of O_2 which represents 1/4 to 2/3 of the total respiration. Since amino acid uptake is not light-dependent in this organism (North and Stephens, unpublished) this intake may represent a useful supplement at night or in light intensities near the compensation point.

These observations on *Platymonas* illustrate the utility of a rather roundabout but necessary rationale for studying the nutritional significance of FAA. In most nutritional investigations the organism is cultured directly on several nitrogen and carbon sources. Conclusions are then drawn concerning the ability of the organism to utilize various substrates. But this approach can be very misleading. Consider the case of *Platymonas* and glutamate (North and Stephens, unpublished). Glutamate is an adequate nitrogen source supporting good growth when supplied a 2×10^{-3} moles/liter in batch culture. When supplied at low concentrations, ^{14}C-glutamate is accumulated slowly by cells from nitrogen-rich culture and is taken up more rapidly by cells on restricted rations. However, when one calculates the rate of growth which could be supported by glutamate at the low concentrations in the habitat, one obtains a doubling time of one and a half to two months at best. Thus glutamate does not appear to be an important nitrogen source for *Platymonas*. Even though an amino acid at high concentration in culture is a competent nitrogen source and can be taken up from low concentration in the medium, it cannot be concluded that it plays a significant role in the economy of the organism under natural conditions. That judgment must be made on the basis of quantitative data extrapolated to natural conditions. Axenic culture at very low substrate levels is not a feasible approach.

DISTRIBUTION OF THE CAPACITY FOR AMINO ACID UPTAKE

All representatives of the soft-bodied phyla of marine invertebrates examined accumulate ^{14}C-amino acids. As just noted, this is not sufficient ground for judgments about the significance of the process but merely illustrates its widespread distribution. In the Chordata, all urochordates and cephalochordates tested show this capacity. In the vertebrates, only hagfishes have been found to remove amino acid from dilute solution. Arthropods were reported to remove organic compounds directly from solution (24) but this appears to be entirely attributable to microorganisms on the exoskeleton (1, 2). Amino acid accumulation appears to be restricted to marine forms in the metazoa. It is difficult to exclude microorganisms as agents so it is not possible to say that no intake of organic material occurs in fresh water forms. However, if it occurs at all, it is very much slower than that observed in marine invertebrates (40).

Table 5 lists the algal divisions in which amino acid uptake

Table 5. *Divisions and phyla of plants and animals in which accumulation of amino acid has been observed*

Division	Phylum
Cyanophyta	Cnidaria
Chrysophyta	Platyhelminthes
Pyrrophyta	Rhynchocoela
Chlorophyta	Ectoprocta
Phaeophyta	Annelida
Rhodophyta	Sipunculoidea
	Echiuroidea
	Mollusca
	Echinodermata
	Hemichordata
	Chordata

has been observed as well as the metazoan phyla examined. In algae, entry of labelled amino acid depends on the previous nutritional history of the cells as well as the particular organism used. Some amino acids are not accumulated under any of our experimental conditions. In other cases, systems for accumulation may become active in regular order as nutrients are restricted. For example, in *Nitzschia*, arginine and lysine are acquired by cells under all growth conditions. However, the capacity to take in glycine, serine, and glutamate from dilute solution appears as nutrient supply is restricted, regardless of the nature of the nitrogen source. Since little work has been done under conditions which can be regarded as favorable for demonstrating amino acid uptake, we can only speculate about the distribution of this ability in phytoplankters.

Macroscopic representatives of the Chlorophyta, Phaeophyta, and Rhodophyta also remove amino acid from dilute solution in some cases (28). In a survey of 20 intertidal and subtidal genera, the capacity for amino acid accumulation was correlated with the occurrence of the alga in areas of pollution or natural contamination. Conversely, the forms which showed little or no capacity for accumulation of amino acids (for example, fleshy red and brown algae) are conspicuously absent from contaminated waters. Insufficient information is available to decide whether the capacity to take in and assimilate dissolved organic material may contribute to the success of macroscopic algae in polluted environments.

REGULATION OF AMINO ACID UPTAKE AND ASSIMILATION - SALINITY

Regulation of amino acid accumulation in euryhaline annelids was studied by Stephens (40). *Nereis limnicola* and *Nereis succinea* survive well in the laboratory at low salinities and are found in estuarine situations in nature. *N. limnicola* can occur in fresh water. Both animals accumulate glycine from dilute solution in sea water. Uptake continues until salinity is decreased to about 1/3 of normal sea water (chlorosity reduced from 520 to 150 mEq/liter). At lower salinities, the animals survive well but can no longer acquire free amino acid from solution. This is also the point where the animal begins osmotic and chloride regulation of the body fluid. Above 150 mEq Cl^-/liter, the organisms are osmoconformers. The cessation of amino acid uptake does not seem to be a response to the decrease in a particular ion in the medium since worms in solutions of galactose (osmotic concentration equivalent to a chlorosity of 200 mEq/liter) can accumulate glycine, albeit at reduced rates.

Several organisms have been examined to determine whether amino acid uptake is influenced by the external concentration of sodium. When sodium is replaced by choline or lithium, uptake rates decrease in the invertebrates tested as well as in the alga, *Platymonas*. However, a more detailed study of the influence of external sodium concentration on amino acid accumulation by the coelomocytes of the annelid, *Glycera*, indicates that the sodium dependence is not sufficient to account for the distribution of FAA between the cells and coelomic fluid (32, 33). As already noted, salinity dependence is not an adequate explanation for the salinity responses of nereid worms.

Salinity is also important in regulating assimilation of FAA. Stephens and Virkar (44) studied the influence of salinity on uptake and assimilation of amino acids by the brittle star, *Ophiactis*. Changes in the concentration of the intracellular FAA pool were also measured. At 50% of normal salinity, the rate of uptake of amino acid drops to about half of control values, FAA pools in the animal decrease to about 50%, but assimilation of amino acid into alcohol insoluble material increases 10-fold. These results support the idea that the FAA pool is regulated by decreased uptake and an increase in rate of amino acid assimilation from the pool into synthetic pathways. This is not true for euryhaline invertebrates generally; the mussel *Mytilus* does not respond to lower salinities by increasing assimilation.

Salinity may also influence rates of efflux. Johannes *et al.* (20) report that loss of FAA from *Bdelloura* is decreased at lower salinities.

REGULATION OF AMINO ACID UPTAKE - AMINO ACID CONCENTRATION

Algae can be induced to accumulate amino acids more rapidly by placing them in low external concentrations of FAA or other nitrogen sources. Alternatively, one might view the low uptake rates in cells grown on high concentrations of FAA as suppression of a normally active system, since the low concentrations are ecologically normal. A similar suppression of amino acid uptake can be demonstrated in sedimentary invertebrates (48, 49). In *Stauronereis*, uptake rates are reduced when the animals are exposed to a 10^{-2} molar solution of an amino acid. Rates may decrease to 11% of control values over a period of 8 to 16 hours. The effect is reversible. Normal rates are re-established 72 hours after the worms are transferred to fresh sea water. This suppression is specific. Incubation of the animals in high concentrations of one amino acid suppresses the rate of intake of relate compounds. The neutral amino acids, dicarboxylic amino acids, and polybasic amino acids interact within each group but do not interact from one group to another.

Wong (48) investigated changes in FAA pools produced by incubation in solutions of amino acids at 10^{-2} moles/liter. The total FAA concentration in the worms did not increase. However, the particular amino acid employed in the medium increased dramatically in internal concentration in some cases (threonine) but not at all in others (glycine). Glycine and threonine are mutual suppressors and one might expect them to behave similarly. The increase in threonine under this treatment might suggest a simple feedback control of influx. However, the same amino acid, threonine, responds as strongly to glycine suppression in the absence of any change in concentration of either glycine or threonine. Therefore such an explanation is inadequate. FAA pools were analyzed from the same population of worms in two consecutive years. There was no significant difference in total FAA but differences in individual amino acids were greater than any which could be induced in the laboratory. This suggests that the regulation of total concentration of FAA is more precise than the regulation of the level of individual amino acids.

Efflux is also influenced by high concentrations of amino acid. Leakage rate of cycloleucine was increased in the presence of 10^{-2} molar glycine and arginine (Table 2). Glycine is a competitive inhibitor of cycloleucine uptake but arginine does not influence cycloleucine accumulation. Thus the effect appears to be nonspecific.

Both the decrease in influx and the increase in efflux in annelids appear at concentrations considerably in excess of levels in the normal environment. In *Platymonas*, suppression of influx occurs at much lower ambient concentrations with perceptible effects at 10^{-4} moles/liter. However, these are also concentrations far in excess of normal. Thus there appears to be a correlation between normal FAA levels in the environment, K_m's for influx, and concentrations at which the systems mediating uptake are suppressed by external amino acid; *i.e.*, an organism from a habitat low in FAA has a low K_m and uptake is suppressed at low ambient concentrations. No increase in efflux in *Platymonas* can be demonstrated even at very high external concentrations of amino acid (Stephens, unpublished).

CONCLUSIONS

Marine habitats are diverse with respect to concentrations of free amino acids as well as total dissolved organic material. The significance of environmental free amino acids must be evaluated in terms of the accumulation capacity of the organism and the normal levels of free amino acid available to it.

Evidence for a nutritive role of external FAA for marine polychaetes is reviewed. Influx is rapid with K_m's roughly ten times higher than natural external concentrations. Efflux is relatively independent of amino acid concentration in the normal ecological range. Influx exceeds efflux at FAA levels which apparently characterize the microhabitat of these organisms. Calculations comparing uptake rate with the respiratory requirements of the worms suggest that uptake of free amino acid is a significant supplement to other feeding pathways. The role of dissolved amino acids in the nutrition of the phytoplankter, *Platymonas*, is also discussed. Influx is rapid. Again, K_m's are about ten times higher than external concentrations in the normal environment. Efflux is small compared with influx at normal ecological levels of FAA. Calculations based on comparing the rate of acquisition of amino acid with nitrogen requirements for growth and with respiratory requirements suggest that uptake of amino acid is a significant nitrogen source and may be a supplementary carbon source for these organisms under some conditions

There are several environmental factors which influence influx and efflux of amino acid and its assimilation by the organism. These include osmotic concentration of the medium, concentration of specifi ions, and concentration of amino acids. Internal factors are also in volved. An example is the relation between the concentration of an amino acid in the FAA pool of the organism and its efflux rate. Thou these may be elements in a control system regulating the intake and use of external amino acid, we do not have sufficient information to give an adequate description of such regulation at present.

A survey of the distribution of the capacity for acquiring FAA from dilute solution indicates that this ability is very widespread among marine organisms. The general occurrence of the phenomenon together with the evidence for participation of external FAA in the economy of those organisms which have been investigated in some detail indicates that continued study of the acquisition and utilization of dissolved organic material by marine organisms is desirable.

SUMMARY

Marine environments differ with respect to dissolved organic materials and free amino acid (FAA) levels. Thus an analysis of the potential significance of amino acid uptake must consider both the capacity of the organism to accumulate these compounds and the availability of FAA in its habitat. Evidence is presented for a nutritional role of FAA in organisms from two quite different marine habitats. Sediment dwelling polychaete worms can obtain amounts of amino acid from the FAA of interstitial water which are comparable to their total respiratory requirements. The green flagellate, *Platymonas*, is planktonic. Although FAA levels are much lower in surface waters, these organisms can accumulate amounts of amino acid which are significant supplements to their nitrogen requirements as well as their respiratory needs.

The influence of external osmotic concentration, concentration of specific ions, and amino acid concentration on influx, efflux, and assimilation of amino acids by marine organisms is reviewed. The capacity to accumulate amino acid from dilute solution is very broadly distributed. In most cases, we have insufficient information to assess its importance to the organism. However, the general occurrence of the phenomenon together with the evidence for its importance in those selected forms which have been studied intensively indicate that the exchanges between marine organisms and the dissolved organic material of their environment deserve careful attention.

REFERENCES

1. Anderson, J.W. Ph.D. thesis, University of California, Irvine, 48 pp., 1969.
2. Anderson, J.W. and G.C. Stephens. *Marine Biology* 4:243, 1969.
3. Bohling, H. *Marine Biology* 6:213, 1970.
4. Chapman, G. and A.H. Taylor. *Nature* 217:763, 1968.
5. Chau, Y.K. and J.P. Riley. *Deep-Sea Res.* 13:1115, 1966.
6. Degens, E.T. *Woods Hole Oceanogr. Inst. Ref. No. 6-52*, Woods Hole, Mass. 27 pp., 1968.
7. Duerr, F.G. Ph.D. thesis, University of Minnesota, 51 pp., 1965
8. Duursma, E.K. *Netherlands J. Sea Res.* 1:1, 1961.
9. Duursma, E.K. In: *Chemical Oceanography*, edited by J.P. Riley and G. Skirrow. New York: Academic Press, 1965. p. 433.
10. Fencl, Z. In: *Theoretical and Methodological Basis of Continuo Culture of Microorganisms*, edited by I. Malek and Z. Fencl. New York, Academic Press, 1966. p. 134.
11. Ferguson, J.C. *Biol. Bull.* 132:161, 1967.
12. Ferguson, J.C. *Biol. Bull.* 133:317, 1967.
13. Ferguson, J.C. *Biol. Bull.* 138:14, 1970.
14. Fogg, G.E. In: *Biochemistry and Physiology of Algae*, edited by R.A. Lewin. New York: Academic Press, 1963. p. 475.

15. Gorkin, R.A. *Physiologist* 13:210, 1970.
16. Hammen, C.S., H.F. Miller, Jr. and W.H. Geer. *Comp. Biochem. Physiol.* 17:1199, 1966.
17. Hellebust, J.A. *Limnol. Oceanogr.* 10:192, 1965.
18. Hellebust, J.A. and R.R.L. Guillard. *J. Phycol.* 3:132, 1967.
19. Jeffrey, L.M. Ph.D. thesis, Texas A & M University, 152 pp., 1969.
20. Johannes, R.E., S.J. Coward and K.L. Webb. *Comp. Biochem. Physiol.* 29:283, 1969.
21. Kerr, N. and G.C. Stephens. *Nature* 194:1094, 1962.
22. Little, C. and B.L. Gupta. *Nature* 218:873, 1968.
23. Mangum, C.P. *Comp. Biochem. Physiol.* 11:239, 1964.
24. McWhinnie, M.A. and R. Johanneck. *Antarctic J.U.S.* 1:210, 1966.
25. Menzel, D.W. *Deep-Sea Res.* 14:229, 1967.
26. North, B.B. and G.C. Stephens. *Biol. Bull.* 133:391, 1967.
27. North, B.B. and G.C. Stephens. In: *Proc. VIth Intl. Seaweed Symposium*, edited by R. Margalef. Subsecretaria Mercante Marina, 1969. p. 263.
28. North, W.J., G.C. Stephens and B.B. North. In press: FAO technical conference on marine pollution, Rome.
29. Pequignat, C.E. *Forma et Functio* 2:121, 1970.
30. Pequignat, C.E. and J.-P. Pujol. *Bull. de la Soc. Linnéeae de Normandie, Ser. 10, 9*:209, 1968.
31. Potts, W.T.W. *Biol. Rev.* 42:1, 1967.
32. Preston, R.L. Ph.D. thesis, University of California, Irvine, 101 pp., 1970.

33. Preston, R.L. and G.C. Stephens. *Am. Zool. 9*:1116, 1969.

34. Reish, D.J. and G.C. Stephens. *Marine Biology 3*:352, 1969.

35. Sloan, P.R. and J.D.H. Strickland. *J. Phycol. 2*:29, 1966.

36. Southward, A.J. and E.C. Southward. *Nature 218*:875, 1968.

37. Staddon, B. *J. Exptl. Biol. 36*:566, 1959.

38. Stephens, G.C. *Biol. Bull. 123*:648, 1962.

39. Stephens, G.C. *Comp. Biochem. Physiol. 10*:191, 1963.

40. Stephens, G.C. *Biol. Bull. 126*:150, 1964.

41. Stephens, G.C. In: *Estuaries*, edited by G.H. Lauff. Washington D.C.: AAAS, 1967. p. 367.

42. Stephens, G.C. *Am. Zool. 8*:95, 1968.

43. Stephens, G.C. and R.A. Schinske. *Limnol. and Oceanogr. 6*:175, 1961.

44. Stephens, G.C. and R.A. Virkar. *Biol. Bull. 131*:172, 1966.

45. Taylor, A.G. *Comp. Biochem. Physiol. 29*:243, 1969.

46. Webb, K.L. and R.E. Johannes. *Limnol. Oceanogr. 12*:376, 1967.

47. Webb, K.L. and L. Wood. In: *Automation in Analytical Chemistry Technicon Symposium 1966*. New York: Mediad, 1967. p. 440.

48. Wong, L. M.A. thesis, University of California, Irvine, 62 pp., 1969.

49. Wong, L. and G.C. Stephens. *Am. Zool. 10*:312, 1970.

D-AMINO ACID OXIDASE: RESPONSE TO A STEREOSPECIFIC CHALLENGE

John J. Corrigan

*Department of Life Sciences
Indiana State University
Terre Haute, Indiana 47809*

The discovery by Krebs that D-amino acids are oxidatively deaminated by slices of mammalian kidney (21) led to a series of studies involving the purification of a flavoprotein enzyme with almost absolute specificity towards D-amino acids. Much subsequent work has been devoted to the relationship between the apoenzyme and its coenzyme, flavin adenine dinucleotide (23), and to the initial steps in the deamination reaction (12). For these investigations the most common source of the enzyme has been either ovine or porcine kidney tissue and, since the pioneer work of Krebs (20, 21), substrate specificity studies have been carried out in several laboratories (2, 3, 17). At the time of its discovery D-amino acid oxidase had no known natural substrates and until the work of Snell and his collaborators commencing in 1943 (29), D-amino acids were not believed to serve any biosynthetic function (10). As evidence accumulated for the presence of C-amino acids in the cell wall structures of bacteria (31) and in microbial polypeptide products (4), various hypotheses were advanced to account for D-amino acid oxidase in the hepatic and renal tissues of vertebrates (23) and in a number of invertebrates (3, 9). Krebs found deaminase activity not only in mammals but also in pigeons, tortoises, frogs and trout (19). At present it seems safe to conclude that all metazoan animals contain at least one form of D-amino acid oxidase with broad substrate specificity and Table 1 illustrates several activity profiles. In the past

Table 1. *Specificity of D-amino Acid Oxidases*[a]

Substrate	Hog Kidney[b]	Sheep Kidney[c]	Octopus Liver[d]
D-Alanine	52	34	53
D-Arginine	5	--	--
D-Glutamate	1	0	57
D-Histidine	4	3	62
D-Isoleucine	73	12	37
D-Leucine	38	7	53
D-Phenylalanine	47	14	--
D-Serine	14	22	23

[a]Relative rates of oxidation as a percentage of the most active substrate at 100.

[b]From Greenstein *et al.* (17) D-proline is 100.

[c]From Bender and Krebs (2) D-tyrosine is 100.

[d]From Blaschko and Hawkins (3) D-α-Aminobutyric acid is 100.

decade, reports have appeared concerning the presence of this enzyme in the brain of about half a dozen vertebrates including humans (35,

13, 26, 16, 11) as well as in granular leukocytes (6). Although ther
is a large literature on various aspects of the biochemistry and action of this enzyme, information on its biological significance is
almost nonexistent.

As a contribution towards the latter question, I present experimental data on the time of appearance, organ distribution and substrate specificity of D-amino acid oxidase in developing chick embryos
This work was undertaken as part of a study to identify cellular factors in animal differentiation which correlate with the synthesis of
certain enzymes. The chick was selected as an amniote vertebrate
which develops free from maternal influence. Some of the findings
have been published in a brief communication (7).

The method used to survey D-amino acid oxidase is illustrated
in Fig. 1 and is based on the enzymatic dehydrogenation of D-allohydroxyproline (I, step 1) to Δ^1-pyrroline-4-hydroxy-2-carboxylic
acid (II). This compound is reacted with 1 N H_2SO_4 (step 2) to yield
the dehydration product pyrrole-2-carboxylic acid (III) which is decarboxylated by heating the acidic solution at 70° for 10 min (step
3). The pyrrole (IV) is reacted with p-dimethylaminobenzaldehyde
to produce a red colored derivative which is then measured photometrically at 550 nm (7, 9). This method was developed for use with
insect tissues and proved to be sensitive enough so that small amounts
of chick tissue homogenate could be rapidly assayed. Substrate specificity studies were carried out by a modification of the Conway
diffusion method (27) and the diffused ammonia was quantitated by
the procedure of Reardon et al. (28) which is based on the formation

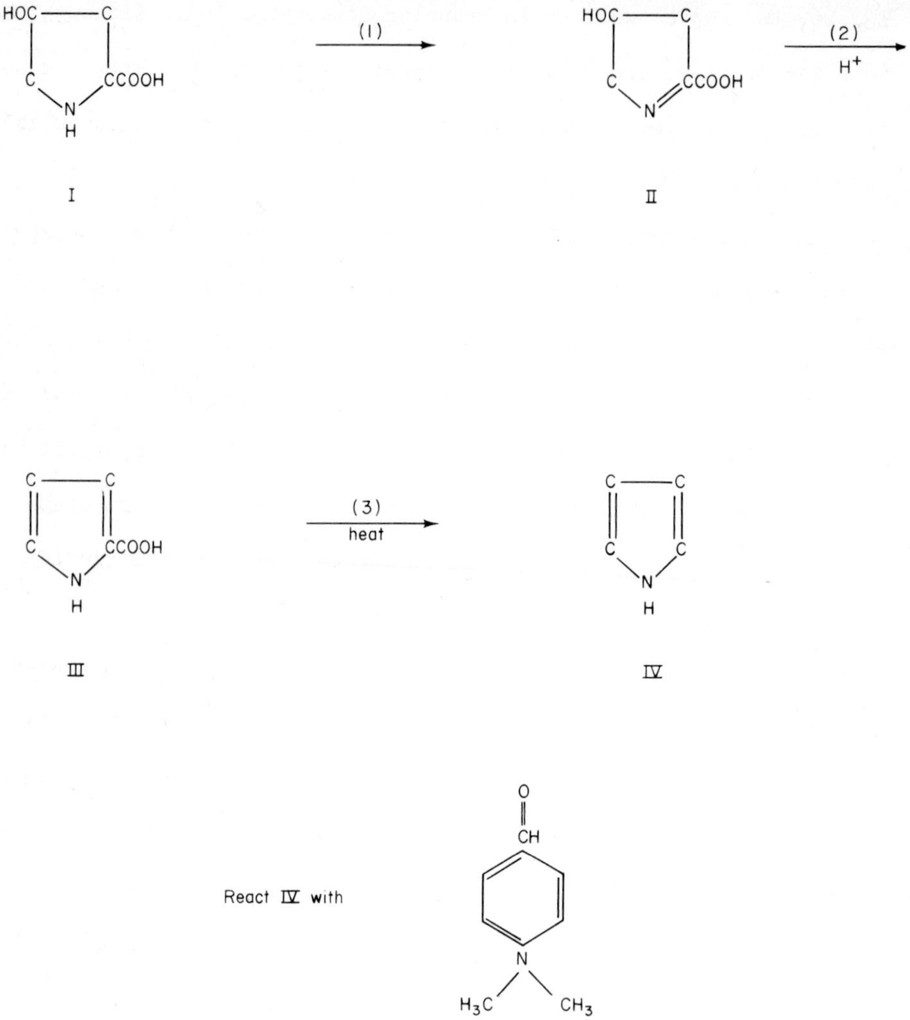

Fig. 1. Conversion of D-allohydroxyproline to pyrrole.

of a colored derivative from ammonia, sodium dichloroisocyanurate and sodium salicylate. The embryos were White Leghorns incubated at 37°. Organ and tissue samples were dissected out from freshly killed embryo and homogenates were prepared for enzyme assay in 0.1 M sodium pyrophosphate buffer (pH 8.3). The age of the embryos was based on days of incubation and also determined according to the Hamburger-Hamilton schedule (18).

RESULTS

The survey of developing tissues in the embryo disclosed that in the hepatic tissue the earliest activity is observed on the seventh day of incubation. This increases several fold on the eighth day then maintains relatively constant specific activity as shown in Table 2. Enzyme activity appears to follow the functional state of

Table 2. *D-Amino acid oxidase in chick embryo liver*

Age of Embryo	Specific Activity[a]
6 day	0
8 day	18
9 day	14
12 day	20
14 day	15
17	15
3 day post-hatch	8

[a]nmoles product/mg protein/hr.

the liver rather closely and the initial rise in activity is correlated with an increase in the total liver protein. Renal tissue activity is shown in Table 3. Detectable enzymatic activity was observed on the seventh day of incubation (Hamburger-Hamilton Stage 30) and by the eighth day this increased about four-fold as the total

Table 3. *D-Amino acid oxidase in chick embryo mesonephros*

Age of Embryo	Specific Activity[a]
6 day	0
8 day	24
9 day	120
12 day	300
14 day	300
17 day	180
Metanephros	
12 day	0
14 day	30
17 day	60
3 day post-hatch	30

[a] nmoles product/mg/hr.

protein doubled. The specific activity in nmoles product synthesized per mg per hr reached a maximum after the tenth day and began to decline as the mesonephros decreased in size. However, even on the eighteenth day, when this organ had atrophied to a very small piece of tissue, enzyme activity was still present. The earliest enzymatic activity in the metanephros appears on the thirteenth day of embryonic life (Stages 37 - 38) and reaches a peak in specific activity around the time of hatch, then declines to a lower equilibrium level.

Experiments were performed to study the fate of D-amino acid oxidase in pieces of mesonephros transplanted onto the chorioallantoi membrane of host embryos. As the morphology of the mesonephric tissue degenerated, enzyme activity also disappeared and this occurred earlier than it does when the mesonephric tissue is *in situ*. These studies together with the evidence for a temporal relationship between organ differentiation and enzymatic activity imply that D-amino acid oxidase is an excellent marker for the onset of morphogenesis in renal and hepatic tissues. During the transplantation experiments it was surprising to find substantial amounts of enzymatic activity in the extra-embryonic membranes but not in the chorioallantoic portion (7). Table 4 shows the specific activity of a series of extra-embryonic tissue samples. It is noteworthy that the earliest D-amino acid oxidase activity occurs in the 1.5 day blastoderm (Stage 9) about 5 to 6 days prior to the first appearance of enzyme in the hepatic and renal tissues. The polar distribution of enzymatic activity on day 3 was particularly intriguing and will be discussed later. By the fifth day, most of the enzyme is in the area *opaca vasculosa* and thereafter becomes associated with the splanchnopleural membranes which form the yolk sac. The specific activity drifts upward to a maximum just prior to the time of hatch. It is interesting that enzyme activity persists in the post-hatch yolk sac while the sac is undergoing atrophy. Enzyme has been detected in yolk sac tissue from a 7-day old post-hatch male and the fluid contents of that sac were devoid of activity. Table 5 shows the results of an experiment to determine if the D-amino acid oxidase of yolk sac is inhibited by sodium benzoate as are the mammalian enzymes (23). One-tenth as much benzoate as substrate resulted in inhibition of 75% of the oxidation

Table 4. *D-Amino acid oxidase in extra-embryonic membranes of chick*

Age of Embryo	Specific Activity[c]
1.5 day[a]	3
3 day[b]	
cephalic half	6
caudal half	3
5 day *a.o. vasculosa*	20
a.o. vitellina	2
6 day	10
8 day	10
12 day	25
14 day	50
16 day	100
18 day	50
20 day	60
5 day post-hatch yolk sac	30
15 day chorioallantois and amniotic sac	0

[a]Hamburger-Hamilton, stage 9.

[b]Hamburger-Hamilton, stage 17.

[c]nmoles pyrrole/mg protein/hr at 37°.

of D-allohydroxyproline compared to the control, and equimolar amounts of benzoate and substrate resulted in complete inhibition.

As mentioned above, several laboratories have reported D-amino acid oxidase in brain tissues including Edlbacher and Wiss (14), Yagi and collaborators (35), Dunn and Perkoff (13), Neims and co-workers

Table 5. *Inhibition of chick yolk sac membrane D-amino acid oxidase by sodium benzoate*

Incubation	Product[b]
Complete[a]	80
Complete with 1 μmole benzoate	20
Complete with 10 μmoles benzoate	1

[a]10 μmoles D-allohydroxyproline, enzyme, FAD, catalase, pyrophosphate buffer pH 8.3, 2 hr at 37°.

[b]nmoles product.

(26), Goldstein (16) and Marchi and Johnston (11). These observations prompted an examination of the chick brain. The brain was divided into two sections, one comprising most of the tissue mass (*corpus striatum* and optic lobes) and representing the anterior region and the other section representing the mid and posterior region (cerebellum plus metencephalon). Table 6 shows that there is somewhat higher activity in the posterior region containing the cerebellum.

Each tissue surveyed for maximum activity was then reacted with eight D-amino acids and glycine in order to determine the substrate specificities. Table 7 shows the data for the liver of 14 and 15 day old chick embryos. The major activities are those of D-alanine, phenylalanine, isoleucine and serine. When the mesonephros was assayed in the same way (Table 8), a similar profile was observed except that the specific activity was about seven times greater than that observed in the liver, consistent with the much higher specific

Table 6. *D-Amino acid oxidase in chick embryo brain*[a]

Age of Embryo	Corpus striatum plus optic lobes	Cerebellum plus metencephalon
13 day	--	10
14 day	6	6
16 day	10	18
17 day	7	10
21 day	5	12
5 day post-hatch	18	24
11 day post-hatch	10	10

[a]Incubated 3 hr at 37° with 50 μmoles of D-allohydroxyproline in each tube. Activity is in nmoles/mg protein.

Table 7. *D-Amino acid oxidase in chick embryo liver*[a]

	14-Day	15-Day
D-Alanine	6.3	4.3
D-Arginine	0.0	0.7
D-Glutamate 2.5 μmoles	0.0	0.0
Glycine	1.0	--
Glycine 100 μmoles	3.0	0.9
D-Histidine	0.0	1.5
D-Isoleucine	3.0	3.4
D-Leucine	0.0	2.9
D-Phenylalanine	9.4	5.0
D-Serine	3.0	2.2

[a]Values are μmoles ammonia produced/mg protein/hr times 100. Concentration of substrate was 10 μmoles per experiment (1×10^{-2}M) unless specified.

Table 8. *D-Amino acid oxidase in chick embryo mesonephros*[a]

	12-Days Old
D-Alanine	36.2
D-Arginine	6.6
D-Glutamate 2.5 μmoles	4.3
Glycine	--
Glycine 100 μmoles	4.9
D-Histidine	7.9
D-Isoleucine	25.9
D-Leucine	16.9
D-Phenylalanine	37.0
D-Serine	16.9

[a]Values are μmoles ammonia produced/mg protein/hr times 100. Concentration of substrate was 10 μmoles per experiment (1×10^{-2}M) unless specified.

activity observed in the experiments using D-allohydroxyproline. The metanephros also showed an activity profile similar to that of the mesonephros except for somewhat lower specific activity (Table 9). The yolk sac gave results which again were similar to the other three tissues (Table 10). The brin showed a much lower although measurable activity for most of the substrates used and the hind portion was noticeably more active than the anterior portion (Table 11). For comparative purposes, a series of L-amino acids was assayed in similar fashion with homogenates from the yolk sac, mesonephros and liver (Table 12). The only amino acid deaminated to a significant extent

Table 9. *D-Amino acid oxidase in chick embryo metanephros*[a]

	14-Days Old	16-Days Old
D-Alanine	29.0	18.4
D-Arginine	4.5	2.7
D-Glutamate 2.5 μmoles	4.0	2.0
Glycine 100 μmoles	--	0.5
D-Histidine	0.0	2.9
D-Isoleucine	1.0	12.8
D-Leucine	11.0	7.0
D-Phenylalanine	24.0	15.5
D-Serine	6.0	7.2

[a]Values are μmoles ammonia produced/mg protein/hr times 100. Concentration of substrate was 10 μmoles per experiment (1 X 10^{-2} M) unless specified.

Table 10. *D-Amino acid oxidase in chick embryo yolk sac*[a]

	13-Day	14-Day
D-Alanine	12.2	11.6
D-Arginine	3.6	1.5
D-Glutamate 2.5 μmoles	1.3	0.8
Glycine	1.3	--
D-Histidine	0.0	3.3
D-Isoleucine	11.0	8.5
D-Leucine	4.7	5.1
D-Phenylalanine	23.0	12.7
D-Serine	11.0	4.1

[a]As in Table 9.

Table 11. *D-Amino acid oxidase in chick brain*[a]

	4-Day Chick	
	C. *striatum* and Opt. Lobes	Cerebellum and Meten.
D-Alanine	0.0	0.5
D-Arginine	0.0	0.5
D-Glutamate	0.0	2.1
Glycine 100 μmoles	0.4	3.0
D-Histidine	0.9	2.8
D-Isoleucine	0.0	2.3
D-Leucine	0.0	2.5
D-Lysine	0.3	3.0
D-Phenylalanine	0.3	2.5
D-Serine	0.1	1.7

[a]Values are μmoles ammonia produced/mg protein/hr times 100. Concentration of substrate was 10 μmoles per experiment (1×10^{-2} M) unless specified.

was L-histidine in the homogenate of liver. This activity may be due to a basic L-amino acid oxidase similar to one which has been reported in the hepatic tissue of turkeys (5).

Although the specific activity of the mesonephros was the highest of all the tissues containing D-amino acid oxidase, because of the different quantities of each embryonic tissue, it became important to compare each organ pooled from several embryos to determine the relative total amounts of enzymatic activity. When this was done using D-allohydroxyproline as the substrate, it became apparent that

Table 12. *L-Amino acid deamination in chick embryo tissues*[a]

	Yolk Sac 14-Day	Mesonephros 14-Day	Liver 14-Day
L-Alanine	0.0	1.6	0.5
L-Arginine	0.0	0.0	0.0
L-Glutamate 2.5 μmoles	0.0	0.0	2.0
Glycine 100 μmoles	0.0	0.0	3.6
L-Histidine	0.0	0.0	33.4
L-Isoleucine	0.0	2.8	0.5
L-Leucine	0.0	0.0	0.0
L-Lysine	0.0	--	0.0
L-Phenylalanine	0.0	0.0	0.0
L-Serine	0.0	0.0	0.0

[a] μmoles of ammonia released/mg protein/hr. Concentration of substrate was 10 μmoles per experiment (1×10^{-2} M) unless specified.

in 12 to 14 day old embryos, the yolk sac accounted for about 80% of all the activity (Table 13). It can be seen that the yolk sac contained nine times more total activity than the liver, the second most active organ of the four compared.

DISCUSSION

From the evidence of early, widespread D-amino acid oxidase activity in avian embryos, it can be inferred that the genes which control this enzyme are selectively repressed only after somatic tissue development is well under way. Consequently, this enzyme seems to be

Table 13. *Relative amounts of D-amino acid oxidase in chick embryo tissues*

Organ	Total Protein[a]	Specific Activity of Enzyme	Total Enzymatic Activity
Yolk sac	37 mg	50 nmoles	1,850 nmoles
Liver	3.32 mg	65	210
Metanephros	0.63 mg	195	123
Mesonephros	0.40 mg	370	128

[a]Determined on 1000 X g supernatant fluid from homogenates of each tissue; 12 to 14-day-old chick embryos were used. Specific activity is in nmoles D-allohydroxyproline oxidized/mg protein/hr.

more important dur ng embryonic life than after the hatch. The yolk itself contains no enzymatic activity but from the earliest time assayed, sizeable quantities of activity are found in the blastoderm. The mesonephros forming area is localized in the cephalic half of the blastoderm at the anterior end of the primitve streak (22). After the mesonephros begins to differentiate, primordial germ cells localized along the peripheral margin of the anterior edge of the blastoderm migrate to the germinal epithelium (genital ridge) which then moves to its definitive position along the lower edge of the mesonephros. It is because of these events that the polar distribution of enzyme activity observed in the third day embryo (Table 4) is especially interesting. After the third day, there is a wider dissemination of enzyme activity throughout the differentiating splanchnopleural membrane system. Whether this is due to progressive derepression or to the migration of D-amino acid oxidase containing cells into the proliferating membranes is unknown at present. A question of some interest is whether or not the primordial germ cells ever contain D-amino acid oxidase. No activity was detected in differentiated gonadal tissue from either male or female embryos.

All the experiments comparing each tissue against the same substrates indicate that each enzyme is similar to the others and to the mammalian enzymes, at least from a catalytic point of view. This observation reflects back on the question concerning the factors which have operated to conserve D-amino acid oxidase in metazoan cells. Evidence for similar enzymes in invertebrates and lower vertebrates (19, 3 9, 16) necessitates an explanation for the broad phylogenetic distribution, whether the underlying mechanism is phylogenetic radiation of the same protein or convergence processes resulting in functionally similar proteins. Although in 1933 there was very little evidence for the occurrence of D-amino acids in organisms, we now know that these substances are relatively common in microorganisms (31, 4) and in animals, particularly among the insects (10). Yet, the majority of D-amino acids are no more toxic than the L-enantiomers and if they are not deaminated to α-keto acids, they are eliminated in urine. The outstanding exception is the anomalous toxicity of D-serine which has been repeatedly demonstrated with microorganisms, plants, invertebrates and vertebrates (10). Since the feeding experiments of Fishman and Artom (15) it has been recognized that D-serine is quite toxic to rats. In subsequent investigations from Fishman's laboratory (24) and from that of Wachstein (32) the kidney was identified as one of the prime targets for the lesions produced by D-serine. More recently, Wise and Elwyn (34) demonstrated in rats that as little as 5 mg of D-serine per 100 g of body weight resulted in generalized hyperaminoaciduria with various amino acids reaching concentrations from ten to forty times above the normal values. The pattern resembles a pathological conditions characterized by impaired renal reabsorption known as Fanconi Syndrome (Table 14).

Table 14. *Amino acid excretion in urine of rats injected with D- or L-serine*[a]

Amino Acid	L-Serine Injection	D-Serine Injection
Alanine	4	123
Aspartic acid	2	21
Glutamic acid	3	122
Glycine	5	110
Histidine	1	20
Isoleucine	1	22
Leucine	1	38
Lysine	3	84
Methionine	1	13
Phenylalanine	1	23
Proline	1	63
Threonine	2	63
Tyrosine	1	27
Valine	1	46

[a] μmoles excreted/day/rat. The urines were pooled from 2 rats injected with either 50 mg D- or L-serine per 100 g of body weight. From Wise and Elwyn (34).

The D-isomer of serine is rarely found as a constituent of microbial metabolism and the preceding observations would be academic were it not for recent investigations which show that D-serine is relatively common in certain invertebrate phyla. In 1959, Beatty, Magrath and Ennor reported the presence of free D-serine in the earthworm *Lumbricus terrestris* together with lombricine and serine ethanolamine phosphodiester, both of which contain a D-serine residue (1). In 1962, Meister, Srinivasan and I found large quantities of DL-serine in the blood and tissues of the silkworm, *Bombyx mori* (30) and in subsequent investigations in my laboratory a number of other species of lepidoptera were shown to contain racemic serine. Reviews of this subject have been published elsewhere (8, 10). To these observations must be added the fact that one of the most common traits of birds is their insectivorousness and this group includes over half of the known species. Even birds which are seed eating as adults feed their young on insects and annelids. Accordingly, nutritional intake of D-serine during ingestion of invertebrates may be the principle factor behind the persistence of D-amino oxidase in animal tissues. On this basis, the early appearance of the enzyme serves to protect the vulnerable embryo from possible endogenous D-serine as well as prepare it for insectivorous nutrition after hatching. All of the enzymes studied in the chick showed high activity towards D-serine consistent with findings on other species, just as the high activity observed for D-alanine, D-phenylalanine, D-isoleucine and D-leucine in chick tissues is typical of other vertebrate preparations (Table 1). Dixon and Kleppe (12) have obtained evidence that there is an alkyl-group binding site on the enzyme which probably accounts for the high reactivity of the more hydrophobic substrates. The slow reaction of glycine, observed in chick embryos, is consistent with evidence that D-amino acid oxidase and glycine oxidase are identical proteins (25).

The presence of D-amino acid oxidase in the brain can also be attributed to the D-serine detoxication hypothesis but there are certain observations which are not explicable in this manner. Goldstein (16) found, while assaying the brain tissue of various vertebrates for this enzyme, that fish have a diffuse distribution of activity, while in rats and mice most of the enzyme is localized in the cerebellum. This observation was extended to cats by De Marchi and Johnston (11) and to human brain tissue by Neims and co-workers (26) who found that adults had about five times more cerebellar enzyme than children. The basis for this difference is unknown.

CONCLUSIONS

The invertebrates evolved long before the vertebrates and are an important source of food for many of the latter. Since a number of annelids and insects contain D-serine, this toxic amino acid is probably a prime factor in the evolutionary process which resulted in selection and conservation of the genes responsible for the synthesis of D-amino acid oxidase. After four decades, we may be close to understanding the physiological role of this unique enzyme and the reasons for its broad phylogenetic distribution.

SUMMARY

In 1933, Hans Krebs discovered that animal tissues oxidatively deaminate D-amino acids. He subsequently showed that the reaction was catalyzed by D-amino acid oxidase, a soluble enzyme widespread in the renal and hepatic tissues of vertebrates. At that time D-amino acids were unknown from biological sources and the enzyme came to be regarded as a biochemical enigma. Further investigations on D-amino acid oxidase were concerned with the substrate specificity and mechanism of action of the enzyme and these studies were confined to preparations purified from hog or sheep kidney tissues. In order to define the relationship of this enzyme to vertebrate development, a study of its specificity and properties was carried out in the organ of embryonic chicks.

D-amino acid oxidase was detected in tissue extracts with a specific, sensitive method involving D-allohydroxyproline as the substrat By this means enzymatic activity was found in the mesonephric, metanephric and hepatic tissues at times closely correlated with the earliest functional state of the respective organs. In addition, a new source of enzyme was discovered in the extra-embryonic membranes, particularly those which form the yolk sac in later embryonic life. This activity appeared on the second day of embryonic development, five day prior to the earliest activity in the renal and hepatic tissues.

All sources of D-amino acid oxidase showed similar substrate-activity profiles characterized by high activity towards D-alanine, phenylalanine, isoleucine, leucine and serine. The fact that all tissues oxidatively deaminate D-serine is of particular interest since this amino acid has repeatedly been observed to be toxic to animals. This toxicity is specific for serine and its basis had not yet been explained.

Several years ago, D-serine was discovered in a number of insects of the order leptidoptera as well as in annelid worms. These studies were the first to disclose abundant natural sources of this amino acid. Because of these findings it is now possible to account for D-amino acid oxidase as an agent which destroys D-serine encountered in dietary sources. Insects and annelids comprise a major portion of the dietary intake of many vertebrates particularly avian species. Accordingly, destruction of D-serine is proposed as the chief function of D-amino acid oxidase in animals and as the main factor conserving this enzyme during vertebrate evolution.

ACKNOWLEDGEMENTS

I thank Mrs. Barbara M. Houlihan, Mrs. Julia Lin and Mr. Michael Kennedy for their skillful experimental assistance. The support of NIH and NSF is gratefully acknowledged.

REFERENCES

1. Beatty, I.M., D.I. Magrath and A.H. Ennor. *Nature 183*:591, 1959.
2. Bender, A.E. and H.A. Krebs. *Biochem. J. 46*:210, 1950.
3. Blaschko, H. and J. Hawkins. *Biochem. J. 52*:306, 1952.
4. Bodanszky, M. and D. Perlman. *Science 163*:352, 1969.
5. Boulanger, P. and J. Bertrand and R. Osteux. *Biochem. Biophys. Acta 26*:]43, 1957.
6. Cline, M.J. and R.I. Lehrer. *Proc. Nat. Acad. Sci. U.S. 62*:756, 1969.
7. Corrigan, J.J. and B.M. Houlihan. *Biochem. Biophys. Res. Commun. 37*:970, 1969.

8. Corrigan, J. In: *Comparative Biochemistry of Nitrogen Metabolism*, Vol. I. *The Invertebrates*, edited by J.W. Campbell. New York and London: Academic Press, 1970. p. 387.
9. Corrigan, J.J., D. Wellner and A. Meister. *Biochem. Biophys. Acta 73*:50, 1963.
10. Corrigan, J.J. *Science 164*:142, 1969.
11. De Marchi, W.J. and G.A.R. Johnston. *J. Neurochem. 16*:355, 1969
12. Dixon, M. and K. Kleppe. *Biochem. Biophys. Acta 96*:368, 1965.
13. Dunn, J.T. and G.T. Perkoff. *Biochem. Biophys. Acta 73*:327, 196
14. Edlbacher, S. and O. Wiss. *Helv. Chem. Acta 27*:1060, 1944.
15. Fishman, W.H. and C. Artom. *J. Biol. Chem. 145*:345, 1942.
16. Goldstein, D.B. *J. Neurochem. 13*:1011, 1966.
17. Greenstein, J.P., S.M. Birnbaum and M. Otey. *J. Biol. Chem. 204* 307, 1953.
18. Hamburger, V. and H.L. Hamilton. *J. Morphol. 88*:49, 1951.
19. Krebs, H.A. In: *The Enzymes*, Vol. 2, edited by J.B. Sumner and K. Myrbäck. New York: Academic Press, 1951. p. 499.
20. Krebs, H.A. *Biochem. J. 29*:1620, 1935.
21. Krebs, H.A. *Z. Physiol. Chem. 217*:191, 1933.
22. Lillie, F.R. and H.L. Hamilton. *Lillie's Development of the Chick*, 3rd Ed. New York: Holt, Rinehart and Winston, 1952. p. 484.
23. Meister, A. and D. Wellner. In: *The Enzymes*, Vol. 6, 2nd Ed, edited by P.D. Boyer, H.A. Lardy and K. Myrbäck. New York: Academic Press, 1962. p. 193.
24. Morehead, R.P., W.H. Fishman and C. Artom. *Am. J. Pathol. 21*: 803, 1945.

25. Neims, A.H. and L. Hellerman. *J. Biol. Chem. 237*:976, 1962.
26. Neims, A.H., W.D. Zieverink and J. Smilack. *J. Neurochem. 13*: 163, 1966.
27. Obrink, K. *Biochem. J. 59*:134, 1955.
28. Reardon, J., J.A. Foreman and R. Searcy. *Clin. Chim. Acta 14*: 403, 1966.
29. Snell, E.E. and B.M. Guirard. *Proc. Nat. Acad. Sci. U.S. 29*:66, 1943.
30. Srinivasan, N.G., J.J. Corrigan and A. Meister. *J. Biol. Chem. 237*:PC 3844, 1962.
31. Strominger, J.L., K. Izaki, M. Marsuhashi and D.J. Tipper. *Fed. Proc. 26*:9, 1967.
32. Wachstein, M. *Arch. Pathol. 43*:503, 1947.
33. Wallace, G.J. *An Introduction to Ornithology.* New York: Macmillan. 1955.
34. Wise, E.M. and D. Elwyn. *Proc. Soc. Exp. Biol. Med. 121*:982, 1966.
35. Yagi, K., T. Nagatsu and T. Ozawa. *Nature 177*:891, 1956.

FOREIGN NITROGENOUS COMPOUNDS -

THEIR EFFECTS AND METABOLISM

Richard H. Adamson

*Laboratory of Chemical Pathology
National Cancer Institute
National Institutes of Health
Bethesda, Maryland 20014*

and

*The Mount Desert Island Biological
   Laboratory
Salisbury Cove, Maine 04672*

Nitrogen was discovered by Daniel Rutherford in 1772 but Scheele, Cavendish, Priestley and others were studying about the same time, "burnt or dephlogisticated air" as air without oxygen was then called. Nitrogen makes up 78% of the air by volume and the amount of this element in the atmosphere is estimated at more than 4000 billion tons. Nitrogen is generally prepared by liquifaction and fractional distillation. The naturally occurring compounds of nitrogen are Chilean nitrate (sodium nitrate), potassium nitrate[1], calcium nitrate and ammonia, of which Chilean nitrate is by far the greatest source. The presence of large amounts of nitrogen in the atmosphere is due

---

[1]From whence it derived its name in 1823 as potassium nitrate was called *niter*.

to the chemical inactivity of the element. Nitrogen is so inert that Lavoisier named it *azote*, meaning without life, yet its compounds are so active as to be most important in foods, poisons, fertilizers, explosives and indeed to life itself since nitrogen is an important constituent of proteins and nucleic acids found in plants and animals.

Man and animals are exposed to a variety of foreign nitrogen containing compounds both inadvertently and intentionally. The effects of some of these compounds and the breakdown and metabolism of representative foreign nitrogenous compounds occurring in the air, and in the water and compounds and drugs to which man are exposed will be discussed in this paper.

The sources of air pollution are divided in Table 1 into 5 categories: transportation, industry (*i.e.*, emissions of the major industrial polluters - pulp and paper mills, steel mills, petroleum refineries, smelters, and chemical manufacturers), power plants, space heating, and refuse disposal. The largest source of pollution is transportation - which contributes 86 million tons per year - 29 million tons more per year than all other sources combined (2).

The chemistry of air pollutants is summarized in Table 2. Air pollutants can be divided into three main classes, the inorganic gases, the organic gases and the aersols (3). Some of the inorganic gases namely the oxides of nitrogen and ozone and one of the organic gases, peroxyacyl nitrate are discussed in this paper.

Nitric oxide (NO) is a colorless somewhat toxic gas which is formed when combustion takes place at high enough temperature to cause a reaction between the nitrogen and oxygen of the air. In

Table 1. *National sources of major air pollutants (millions of tons per year)*

Source	Carbon Monoxide	Sulfur Oxides	Hydro-Carbons	Nitrogen Oxides	Particulate Matter	Misc. Other	Total
Transportation	66	1	12	6	1	*	86
Industry	2	9	4	2	6	2	25
Power plants	1	12	*	3	3	*	20
Space heating	2	3	1	1	1	*	8
Refuse disposal	1	*	1	*	1	*	4
Total	72	25	18	12	12	4	143

*Less than 1.

Table 2. *Chemistry of air pollutants*

Class	Subclass	Example
Inorganic gases	Oxides of nitrogen	Nitrogen dioxide, nitric oxide
	Oxides of sulfur	Sulfur dioxide, sulfuric acid
	Inorganics	Ozone, ammonia, carbon monoxide
Organic gases	Hydrocarbons	Benzene, methane
	Aldehydes	Formaldehyde
	Other organics	Peroxyacyl nitrates, chlorinated hydrocarbons
Aersols	Solid particulate matter	Dust, smoke
	Liquid particulates	Fumes, oil mists

most cities the autmobile is the largest single source of this compound. Nitrogen dioxide is a further oxidation produce of nitric oxide - it is yellow brown in color, is more toxic than nitric oxide and is the only important and widespread pollutant gas that is colored. As a result it can significantly affect visibility. Nitrogen dioxide can also react with raindrops or water vapor in the air to produce nitric acid (HNO_3) which, even in small concentrations can corrode metal surfaces in the immediate vicinity of the source.

In addition, smog of the Los Angeles type results from the absorption of energy from the sun by nitrogen dioxide in the presence of hydrocarbons (Fig. 1) (10). In this process nitric oxide and

$$NO_2 \xrightarrow[\text{Hydrocarbons}]{\text{Sunlight}} NO + O\cdot$$

$$O_2 + O\cdot \longrightarrow O_3$$

$$O\cdot \text{ or } O_3 + \text{Hydrocarbons} \longrightarrow \text{Peroxyacyl nitrates and aldehydes}$$

Fig. 1. Mechanism of photochemical smog formation.

atomic oxygen are formed. The atomic oxygen reacts with other oxygen molecules and other constituents of auto exhausts to form a variety of products including ozone (10). Ozone is one of the most toxic of air pollutants. It can cause choking, headache, coughing and fatigue as well as damage the leaves of plants and deteriorate fabrics. In addition, ozone is also a participant in a highly complex series of chemical reactions which result in the formation of equally undesirable chemicals. For example, both PAN (peroxyacyl nitrate) and formaldehyde can be formed. These two compounds are powerful irritants of the eyes, skin and respiratory tract and can also damage plant life.

In addition to the nitrogenous compounds in air, man may intentionally expose himself to a variety of foreign nitrogenous compounds (Table 3). Thus man may by smoking expose himself to nicotine and nornicotine. Nicotine may well be mutagenic and nitrosonornicotine resulting from a reaction between NO or NO_2 and nornicotine may well be a potent carcinogen (8). In addition man may expose himself to quinine in a number of ways, to cyclamate and/or saccharin, two known bladder carcinogens in rodents, and to vitamin B_1 and caffeine.

Table 3. *Foreign nitrogenous compounds to which man may be intentionally exposed*

Class	Chemical
Tobacco	Nicotine
"Soft drink"	Tonic water (quinine)
Artificial sweetner	Sodium cyclamate
Artificial sweetner	Saacharin
Vitamins	B_1 (thiamine)
Coffee	Caffeine

A number of our important drugs are nitrogenous containing compounds (Table 4). The barbiturates, the narcotic and antipyretic analgesics, the anticancer agents, nitrogen mustard, and 6-mercaptopurine and the antimicrobial sulfonamides are nitrogen containing. Among other drugs which are nitrogenous compounds are the CNS stimulant amphetamine, the hallucinogen LSD, nitrous oxide, organic nitrates which are vasodilators, the antibiotics, penicillin and streptomycin and two well known transquilizers, chloropromazine and Libriu

There are also numerous nitrogenous compounds to which man may inadvertently be exposed (Table 5). These include two insecticides parathion and imidan. Parathion is one of the most toxic of the insecticides smong agricultural workers (12). For example, in California in 1963, there were 345 cases of occupational poisoning of all types; 267 of these were due to parathion. Two naturally occurring poisons, muscarine and strychnine are nitrogen containing compounds. Man may also be exposed to the rocket fuel hydrazine or to

Table 4. *Widely used drugs which are nitrogenous compounds*

Class of drugs	Drug
Sedative	Barbiturates
Analgesic	Morphine
Antipyretic analgesic	Phenacetin
Antipyretic analgesic	Acetanilid
Anticancer agent	Nitrogen mustard
Anticancer agent	6-Mercaptopurine
Antimicrobial	Sulfonamides
CNS stimulant	Amphetamine
Hallucinogen	LSD
Anesthetic	Nitrous oxide
Antihypertensive	Organic nitrates
Antibiotic	Penicillin
Antibiotic	Streptomycin
Tranquilizer	Chlorpromazine
Tranquilizer	Librium

hydrazine analogs and to a number of aromatic amines and azo compounds some of which are potent carcinogens. For example, the nitrosamines, formerly used as industrial solvents, have produced tumors experimentally from fish to various subhuman primates (11, 14, 17). Some cheeses (Brie, Camembert, New York Cheddar) are rich in tyramine, a compound normally harmless because of its rapid oxidation by monoamine oxidase (MAO). However, in the presence of MAO inhibitors

Table 5. *Foreign nitrogenous compounds to which man may inadvertently be exposed*

Class	Chemical
Insecticide	Parathion
Insecticide	Imidan
Mushroom poison	Muscarine
Poison	Strychnine
Rocket fuel	Hydrazine
Industrial chemicals	Aromatic amines
Dyestuffs and food coloring	Azo compounds
"Food amines"	Tyramine - cheese
Well water	Inorganic nitrate
Pickled meats	Sodium nitrate
Fertilizer	Technical sodium nitrate
Paralytic shellfish poison	$C_{10}H_{17}N_7O_4$

tyramine (present in cheese) has produced hypertensive crises and in a few instances fatal cerebral hemorrhage (7).

Man and animals may be exposed to high amounts of inorganic nitrate which may occur in well water, in pickled meats, in over fertilized vegetables or even in public drinking water. Often fatal methemoglobinemia is seen in infants exposed to water containing inorganic nitrate in concentrations as low as 50 mg/liter. Three factors are responsible; first, the less acid stomach of infants allows invasion of the upper gastrointestinal tract by bacteria which reduce nitrate to nitrite which is absorbed and is the direct cause of methemoglobinemia. Second, infants are less able to reduce methemoglobin back to the ferrous form and third, young infants have considerable amounts of fetal hemoglobin in their blood which is more easily converted to methemoglobin than is the adult form. The paralytic shell fish poison with an empirical formula of $C_{10}H_{17}N_7O_4$ is found in several species including the Alaska butterclam, *Saxidomus giganteus* and the California mussel, *Mytilus californianus*. The toxic compound called saxitoxin apparently contains a perhydropurine nucleus which incorporates two guanidinium moieties. This toxic compound appears to be the same as that of the red tide toxin which will be discussed later (6).

In addition to foreign compounds to which man is exposed both inadvertently and intentionally and to compounds occurring in the air there are various foreign compounds to which man is exposed in his drinking water and to which fresh water and marine species are exposed. Our waterways have been fouled from the Androscoggin in Maine to the Zambezi in Africa. No nation has remained untouched - the Gulf of Mexico, the Irish Sea, the Potomac, the Rhine, the Tiber, the Ural - all have been polluted (5). The types and sources of water pollution are summarized in Table 6. These vary from organic wastes and sewage to metals, pesticides, oil, radioactive pollutants and thermal pollution.

A recent survey has shown that at least one million Americans are drinking water that is potentially hazardous due to chemical or bacterial contamination. This survey of public water covered 969 communities ranging from major urban centers to small towns of less than 5,000 persons. For example, the survey included 278 systems in two California

Table 6. *Types of water pollutants*

Organic wastes

Infectious agents from sewage

Synthetic organic chemicals

Inorganic chemicals

Metals

Pesticides

Herbicides

Oil

Thermal

Mineral substances

Sediments

Radioactive pollutants

Plant nutrients

---

counties, San Bernardino and Riverside (13). In these two counties 26% of the systems exceeded recommended limits and 16% exceeded federal mandatory limits for at least one contaminant. In these two counties 44 systems exceeded recommended limits for dissolved solids, 34 for iron, 28 for bacterial content, 14 for sulfate, and 12 for nitrate. Two were above the mandatory limits for arsenic and seven exceeded the mandatory limits for lead. However, the pollution of our drinking water is minimal compared to the pollution of the waterways used for recreation, sport, and in which fresh and marine water species live. Table 6 summarizes the types of water pollutants and

Table 7 illustrates some of the nitrogenous compounds which fresh

Table 7. *Foreign nitrogenous compounds which fresh water and marine species may be exposed to*

Class	Chemical
Insecticide	Parathion
Weed killer	2-Methyl-4,6-dinitrophenol
Molluscide	Zectran
Fertilizer and wastes	Nitrates, nitrites, and nitrogenous end products
Degradation product of animals and plants	Trimethylamine
Red tide toxin	$C_{10}H_{17}N_7O_4$

water and marine species may be exposed to. Among the compounds are insecticides and weed killers which through run off or direct dumping enter the streams, rivers and lakes; the molluscides, fertilizer and wastes sometimes entering as run off and sometimes discharged as raw sewage into our rivers; degradation products of animals and plants for example, trimethylamine and red tide toxins, a product of the dinoflagellate especially *Gonyaulax catenella* and *Gonyaulax tamarensis*. The dinoflagellate organisms produce toxic metabolite(s) which are liberated into the water causing mass mortalities of fish and other organisms as a result of local blooms. When these blooms are near land strong sea breezes can blow toxic spray inland resulting in respiratory distress for inhabitants of nearby communities (6). As previously alluded to this toxin is the same as that found in various shellfish.

Having discussed some of the foreign nitrogenous compounds to which man and animals are exposed one now wishes to examine the metabolic fate of some of these compounds in mammals and in fresh water and marine species. When a foreign compound enters the body it may undergo spontaneous reactions to give other compounds; it may be excreted unchanged, or it may be metabolized by enzymes; microsomal, non-microsomal, or bacterial. Some of the routes of metabolism of foreign nitrogenous compounds by microsomal enzymes are given in Table 8 (1, 15, 19). These include reduction of the azo bond, an

Table 8. *Pathways of metabolism of foreign nitrogenous compounds by microsomal enzymes*

Pathway	Example
Azo reductase	Prontosil to sulfanilamide
Nitro reductase	p-Nitrobenzoic acid to p-aminobenzoic acid
N-Hydroxylation	2-Acetylaminofluorene to N-hydroxy-2-acetylaminofluorene
N-Oxidation	Trimethylamine to trimethylamine oxide
Oxidative deamination	Amphetamine to phenylacetone
N-Dealkylation	Aminopyrine to 4-aminoantipyrine
N-Glucuronide formation	Aniline to aniline-glucuronide
Mercapturic acid formation	2-Benzothiazolesulfonamide to 2-mercaptobenzothiazole

example being reduction of prontosil to sulfanilamide. Aromatic nitro groups can be reduced to amines by a microsomal nitro reductase. Both

N-hydroxylation and N-oxidation of nitrogen compounds can occur. The carcinogenicity of 2-acetylaminofluorene is thought to be related to the extent of N-hydroxylation. In addition, nitrogen compounds may be deaminated, N-dealkylation may occur, N-glucuronide formation may occur as in the case of aniline or mercapturic acid formation may take place.

Some of the pathways which foreign nitrogenous compounds are metabolized by non-microsomal enzymes are summarized in Table 9 (1, 15, 19). Thus, nitrogenous compounds may undergo hydrolysis by esterases or amidases, they may be acetylated as in the case of sulfanilamide, they may undergo deamination by monoamine or diamine oxidase. In addition, hydroxamic acids may be reduced and various nitrogenous compounds may undergo ribonucleotide formation, N-methylation or transsulfuration.

Generally, the microsomal and non-microsomal enzymes which metabolize foreign nitrogenous compounds are present in most mammalian species and differences among the mammalian species generally are in extent or rate of metabolism rather than presence or absence of these pathways of metabolism.

A number of the microsomal and non-microsomal enzymes that metabolize foreign nitrogenous compounds are also present in fresh water and marine species (1). For example, rhodanese, the enzyme which catalyzes the conversion of cyanide to thiocyanate occurs widely in nature from bacteria to man. It is found in the cod and mullet and frogs are particularly rich in the enzyme. Acylation of sulfanilamide is found in both elasmobranchs and teleosts. Azo reductase is present in sharks, rays and various teleosts (Table 10). Azo reductase

Table 9. Pathway of metabolism of foreign nitrogenous compounds by non-microsomal enzymes

Pathway	Enzyme Involved	Example
Hydrolysis	Esterase	Procaine to p-aminobenzoic acid
Hydrolysis	Amidase	Procaine amide to p-aminobenzoic acid
Acylation	Acetyl-CoA	Sulfanilamide to N_4-acetylfulfanilamide
Deamination	Monoamine oxidase	Tyramine to p-hydroxyphenacetaldehyde
Deamination	Diamind oxidase	$2NH-(CH_2)_n-NH_2$ to $2HN-(CH_2)_{n-1}-CHO$
Reduction of hydroxamic acids	Unknown	Salicylhydroxamic acid to salicyclamide
Ribonucleotide formation	Nucleotide pyrophosphorylase	6-Mercaptopurine to 6-mercaptopurine nucleoside monophosphate
N-Methylation	"Soluble" N-methylases	Nicotine to N-methylnicotine
Transsulfuration	Rhodanese	Cyanide to thiocyanate

Table 10. *Azo-reductase activity in liver of various marine species*

Species	Temp °C	Azo reductase (μmoles sulfanilamide formed/g liver/hr)
Fish		
Lemon shark	26	0.61
Dogfish	37	0.33
Stingray	26	0.76
Skate	37	0.16
Hagfish	37	0.13
Lungfish	37	1.2
Barracuda	26	0.76
Yellow-tail snapper	26	0.78
Amphibia		
Frog (*R. pipiens*)	21	0
Frog (*R. pipiens*)	37	1.34
Frog (*R. catesbeiana*)	21	0.58
Frog (*R. catesbeiana*)	37	1.18
Reptile		
Turtle	21	0.50
Turtle	37	1.44
Mammal		
Mouse (CDBA strain)	37	6.8

activity in fish was generally one-tenth of that in various rodent species. Enzyme activity in the turtle and aquatic amphibia was one-fifth of that found in rodent preparations. Nitro reductase activity was absent in several species of shark but was present in the skate, *Raja erinacea* and in the liver of several fresh water and marine teleosts (Table 11). The enzyme activity in some teleosts was approximately one-fourth of that found in various rodent species (1).

Trimethylamine is a highly volatile basic amine which may be

Table 11. *Nitro-reductase activity in liver of various piscine species*

Fish Species	Temp °C	Nitro reductase (µmoles p-aminobenzoic acid formed/g liver/hr)
Salt water		
Lemon shark	26	0
Dogfish	37	0
Stingray	26	0
Skate	37	0.49
Barracuda	26	0.50
Yellow-tail snapper	26	0.66
Fresh water		
Pacific lamprey	25	0.02
Flathead catfish	25	0.05
Channel catfish	25	0.06
Yellow perch	25	0.07
White crappie	25	0.07
White sturgeon	25	0.08
Bluegill	25	0.09
Northern pike	25	0.14
Silver salmon	25	0.17
Smallmouth bass	25	0.18
Steelhead trout	25	0.18
Rainbow trout	25	0.21
Chinook salmon	25	0.21
Sockeye salmon	25	0.31
Carp	25	0.33
American shad	25	0.48
Coarsecale sucker	25	0.63

produced from choline by the bacterial flora of the gut. Trimethylamine may also be a degradation product of plants and fish. This compound may be metabolized by N-demethylation to form urea and formaldehyde or may be oxidized to trimethylamine-N-oxide (Fig. 2). It is this latter pathway of metabolism that has been studied to a greater extent. The oxidation of trimethylamine to trimethylamine-N-oxide is accomplished by hepatic microsomes and requires NADPH and molecular

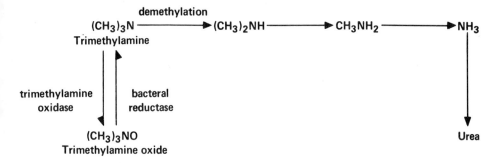

Fig. 2. Metabolism of trimethylamine.

oxygen. This enzyme system is present in mammals (rat, guinea pig, rabbit, sheep and hog), in birds (chicken), in reptiles (turtle), in amphibia (*Necturus* and frog) and in both fresh and salt water fish including some elasmobranchs. However, no invertebrates catalyzed the oxidation of trimethylamine, even zooplankton which contains large quantities of trimethylamine-*N*-oxide were unable to catalyze the oxidation of trimethylamine (4). From the distribution of the hepatic microsomal enzyme that oxidizes trimethylamine and other evidence it is likely that this hepatic microsomal enzyme is concerned with the detoxication of trimethylamine and in species where trimethylamine-*N*-oxide has a role as an end-product of nitrogen metabolism, a methyl donor, or in osmoregulation an alternate synthetic pathway may well be important.

The metabolic fate of five foreign nitrogenous compounds previously mentioned is summarized below. These compounds are amphetamine, cyclamate, nicotine, 2-methyl-4,6-dinitrophenol (DNOC) and parathion. Man exposes himself to the first three compounds and both man and fresh water and marine species may be exposed to DNOC and parathion.

The metabolism of amphetamine by some mammals is summarized in Fig. 3. The major routes of metabolism are hydroxylation to p-hydroxy amphetamine and deamination to benzylmethyl ketone which can be further oxidized to benzoic acid. In man approximately 30% of a dose

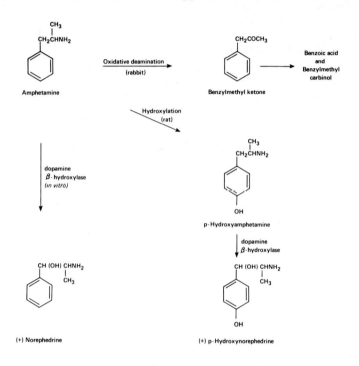

Fig. 3. Metabolism of amphetamine.

of amphetamine is excreted unchanged with the major metabolites being benzylmethyl ketone (3%) benzoic acid (20%) and p-hydroxyamphetamine (3%). Likewise in the dog the amphetamine is excreted mainly unchanged (38%) and as benzoic acid (32%). However, in the rat amphetamine is primarily hydroxylated; approximately 60% of a dose being excreted as p-hydroxyamphetamine. In the rabbit amphetamine is primarily deaminated - 22% of a dose can be identified as benzylmethyl ketone and

27% as benzoic acid. Thus, the extent and type of metabolism of amphetamine by various mammals varies with the species (9).

Cyclamate is a highly polar compound and is unlikely to be extensively metabolized by microsomal or non-microsomal enzymes of various tissues but could be metabolized by bacterial enzymes. One possible reaction is its conversion to cyclohexylamine and inorganic sulfate by hydrolysis. Other possible metabolites of cyclamate are seen in Fig. 4 (16, 18). They include the *N*-acetyl or *N*-methyl derivatives of cyclohexylamine, cyclohexylhydroxylamine, 1-, 2-, 3-, or 4-

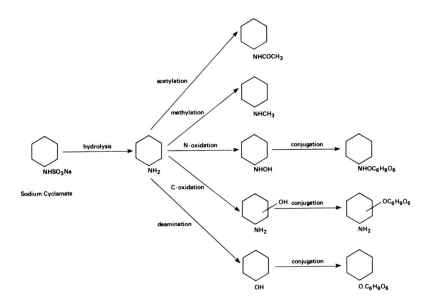

Fig. 4. Some possible metabolites of cyclamate.

hydroxycyclohexylamine, cyclohexanol, and the glucuronic acid conjugates of the last three compounds. When cyclamate is given orally to humans who have been on a cyclamate free diet as expected the sweetner is excreted unchanged partially in the urine and partially in the feces. Other species including the rat also excretes cyclamate unchanged in the urine and feces. However, after daily doses of cyclamate to rats or humans the ability to metabolize cyclamate is acquired apparently by the bacterial flora of the gastrointestinal tract. This ability to metabolize cyclamate depends upon the species, the subject and on a continued intake of cyclamate. Thus humans can convert as much as 17% of a dose of cyclamate to cyclohexylamine after being on a cyclamate diet for as little as 12 days. In addition the cyclohexylamine formed from cyclamate can be further metabolized and indeed the glucuronic acid conjugate of cyclohexylamine is its major metabolite in the rabbit (16).

Nicotine undergoes extensive metabolism by mouse, rat, dog and man (Fig. 5 and Fig. 6). Thus nicotine is metabolized by oxidation

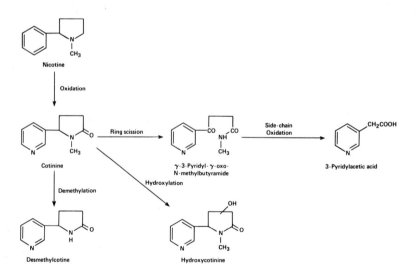

Fig. 5. Metabolism of nicotine.

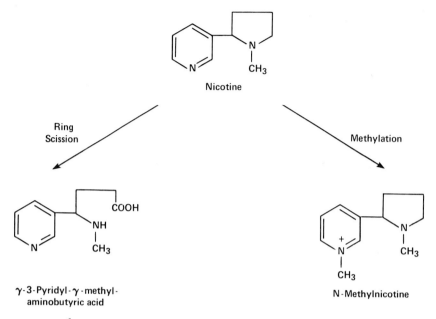

Fig. 6. Metabolism of nicotine.

(Fig. 5) to cotinine which undergoes further metabolism by demethylation, hydroxylation or ring scission. Also, nicotine itself undergoes scission of the pyrrolidine ring to give γ-3-pyridyl-γ-methylaminobutyric acid and is N-methylated to give N-methylnicotine (Fig. 6). Most of these metabolites are excreted in the urine and have been detected in the urine of humans after smoking tobacco or after the accidental ingestion of nicotine (15).

2-Methyl-4,6-dinitrophenol (DNOC) is an effective ovicide, fungicide and weed killer. It is slowly eliminated from the body of various mammals including man (15). Its metabolism by fresh water and marine species which might be exposed to this nitrocresol is unknown. The compound (Fig. 7) may be metabolized by reduction and further metabolized by oxidation to 3-amino-5-nitrosalicylic acid or by acetylation to 6-acetamido-2-methyl-4-nitrophenol a metabolite that is twenty

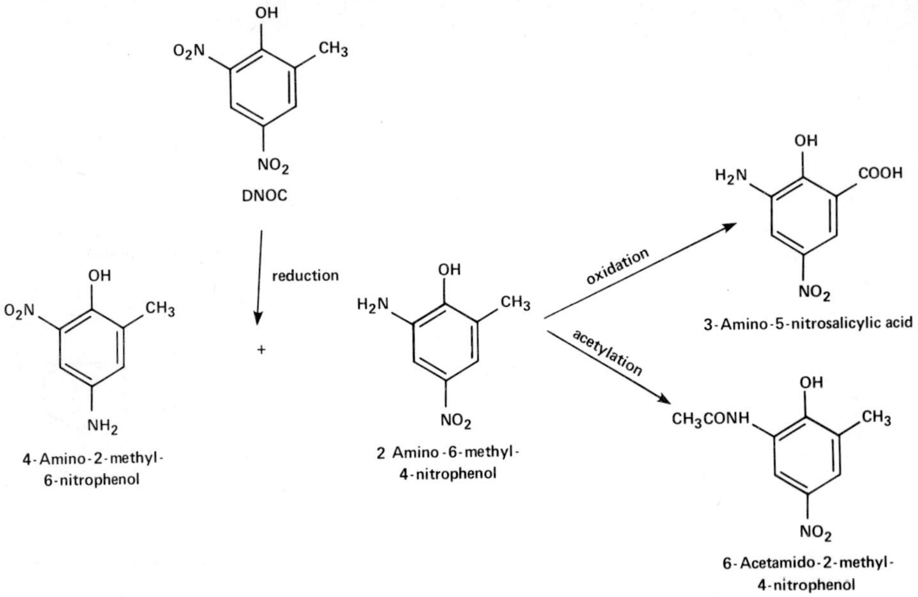

Fig. 7. Metabolism of 2-methyl-4,6-dinitrophenol (DNOC).

times less toxic than the parent compound. In the rabbit the principle metabolite is this less toxic acetylated compound. Other metabolites found in various species are some unchanged parent compound, conjugated DNOC (Fig. 8), 4-amino-2-methyl-6-nitrophenol and 3-amino-5-nitrosalicylic acid (Fig. 7).

The activation and detoxication of parathion is summarized in Fig. 9. The insecticide parathion *per se* has no anticholinesterase activity but is readily metabolized by various species including insects, fish and mammals to its oxygen analog paraoxon, a potent inhibitor of cholinesterase. This activation process in which the P=S group is converted into a P=O group is localized in the hepatic microsomes of fish and mammals and the enzyme system requires NADPH

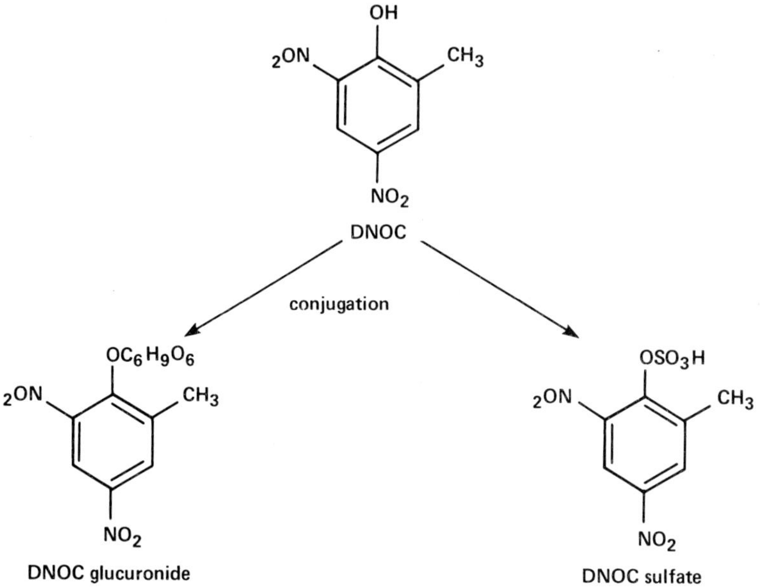

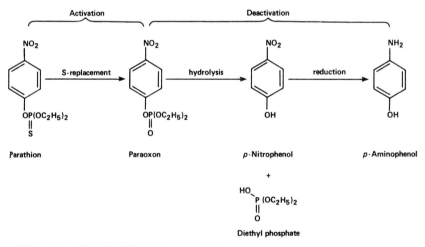

Fig. 8. Metabolism of 2-methyl-4,6-dinitrophenol (DNOC).

Fig. 9. Metabolism of parathion.

and oxygen (1). Paraoxon is further metabolized by hydrolysis to p-nitrophenol and diethyl phosphate, the p-nitrophenol being excreted in the urine as p-aminophenol and its conjugates. The relative susceptibility of various species to poisoning by parathion will be due in part to the rate at which the oxygen analog, paraoxon, accumulates this in turn being dependent on both the formation of paraoxon as well as the rate of destruction by tissue hydrolases or other mechanisms of inactivation.

Thus, foreign nitrogenous compounds may be detoxified or toxic metabolites may be formed by microsomal, non-microsomal or bacterial enzymes of the body.

SUMMARY

Man and other species are exposed both inadvertently and intentionally to a variety of foreign nitrogenous compounds in the air, in the water, and as drugs, pesticides or food additives. The effects of some of these foreign nitrogenous compound and their metabolic fate has been discussed.

ACKNOWLEDGEMENTS

The author is grateful for the help of Miss Bettie S. Braver in the preparation and typing of this manuscript.

REFERENCES

1. Adamson, R.H. *Fed. Proc. 26*:1047, 1967.
2. *Air Pollution Primer*, National Tuberculosis and Respiratory Disease Association, New York, 1969.
3. Altman, P.L. and D.S. Dittmer, eds. *Environmental Biology*. Bethesda, Maryland: Fed. Am. Societies Exp. Biol., 1966.
4. Baker, J.R., A. Struempler and S. Chaykin. *Biochim. Biophys. Acta 71*:58, 1963.

5. Bangor Daily News, July 25 - 26, 1970 and the Boston Sunday Globe, July 26, 1970.
6. Baslow, M.H. *Marine Pharmacology*. Baltimore: The Williams and Wilkins Co., 1969.
7. Blackwell, B. *Lancet* 2:414, 1963.
8. Boyland, E., F.J.C. Roe and J.W. Gorrod. *Nature* 202:1126, 1964.
9. Dring, L.G., R.L. Smith and R.T. Williams. *J. Pharm. Pharmac.* 18:402, 1966.
10. Haagen-Smit, A.J. *Scientific American* 210:25, 1964.
11. Kelly, M.G., R.W. O'Gara, R.H. Adamson, K. Gadekar, C.C. Botkin, W.H. Reese, Jr. and W.T. Kerber. *J. Natl. Cancer Inst.* 36:323, 1966.
12. Kleinman, G.D. *Occupational Disease in California Attributed to Pesticides and other Agricultural Chemicals*, 1963. State of California Department of Public Health, Berkeley, California.
13. Los Angeles Times, August 18, 1970.
14. Magee, P.N. and R. Schoental. *Brit. Med. Bull.* 20:102, 1964.
15. Parke, D.V. *The Biochemistry of Foreign Compounds*. New York: Pergamon Press, 1968.
16. Renwick, A.G. and R.T. Williams. *Proc. First Int. Sugar Research Conference*, The International Sugar Research Foundation, Inc. Bethesda, Maryland, 1970.
17. Stanton, M.F. *J. Natl. Cancer Inst.* 34:117, 1965.
18. Williams, R.T. *Ann. N.Y. Acad. Sci.*, in press.
19. Williams, R.T. *Fed. Proc.* 26:1029, 1967.

EXCRETION OF FOREIGN NITROGENOUS COMPOUNDS[1]

Roy P. Forster

*Department of Biological Sciences
Dartmouth College
Hanover, New Hampshire   03755*

N-OXIDES; TRIMETHYLAMINE OXIDE

Trimethylamine oxide (TMAO) is a detoxication product of the volatile strong base trimethylamine. Trimethylamine, as such, is rapidly eliminated by the elasmobranch, *Squalus*, and presumably by all fish (24). The oxide is of interest to comparative renal physiologists because in some forms such as the aglomerular goosefish, *Lophius*, it is actively secreted, whereas in others such as the elasmobranch, *Squalus*, it is actively reabsorbed by the renal tubules. Here its retention in the body fluids of sharks, skates, rays and chimaeroids plays a role similar to that of urea in rendering them osmotically superior to their marine environment.

TMAO is one of a class of saltlike compounds called amine oxides or N-oxides which are much less basic than the original tertiary

---

[1] The original investigations from the author's laboratory mentioned in this review were supported in part by National Heart and Lung Institute grant HE 04457.

amines. The lone pair of electrons of nitrogen that form the N-O bond have a higher affinity for oxygen than for nitrogen which results in a formal charge, a large dipole moment, and a highly increased polarity (Fig. 1). Bickel (4) and Cheymol et al. (9) have

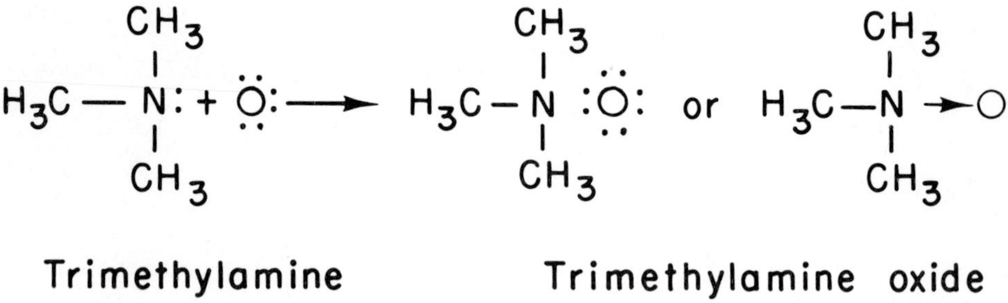

$$H_3C-\underset{\underset{CH_3}{|}}{\overset{\overset{CH_3}{|}}{N:}} + \ddot{\underset{..}{O}}: \longrightarrow H_3C-\underset{\underset{CH_3}{|}}{\overset{\overset{CH_3}{|}}{N}}:\ddot{\underset{..}{O}}: \quad \text{or} \quad H_3C-\underset{\underset{CH_3}{|}}{\overset{\overset{CH_3}{|}}{N}} \rightarrow O$$

Trimethylamine   **Trimethylamine oxide**

Fig. 1. A tertiary amine such as trimethylamine undergoes an interaction with oxygen in which the electron affinity for oxygen is higher than for nitrogen. This bond is frequently represented by an arrow which indicates the direction of electron transfer.

recently reviewed the physico-chemical, physiological and pharmacological properties of TMAO and the amine oxides.

In the marine teleosts where TMAO is actively secreted its transport is inhibited competitively by tetraethylammonium ion and by the basic cyanine dye No. 863 (Table 1). In the marine elasmobranch, *Squalus acanthias*, on the other hand, TMAO is almost completely reabsorbed by the renal tubules (95 - 98%), and its high and steady concentration (80 - 90 µmoles per ml) in the body fluids of this shark accounts for about 20% of the total amount of osmotically active organic solute. The oppositely directed active transport processes in these two species handling the identical compound do not,

Table 1. *Secretion of trimethylamine oxide inhibited competitively by the injection of tetraethylammonium ion into 4.5 kg aglomerular teleost goosefish,* Lophius *(20)*

Urine Sample	Time Hours	Urine flow ml/kg/day	TMAO		TEA	
			µmole/ml	µmole/ kg/day	µmole/ml	µmole/ kg/day
1	0 - 3.4	20.3	20.0	406	-	-
2	3.4- 6.7	22.0	14.5	319	-	-
	7 mg/kg TEA-Br intramuscularly at 6.7 hr					
3	6.7- 9.1	20.2	4.0	81	2.43	49.0
4	9.1-13.7	18.3	1.8	32	1.97	36.1
5	13.7-17.4	18.9	3.3	62	1.18	22.3
6	17.4-20.0	18.6	2.8	51	0.75	14.0

however, appear to be characterized similarly with respect to competitive inhibition. As can be seen in Table 2, doses similar to or higher than those that reduced the excretion of TMAO in *Lophius* by 85% had absolutely no effect on the reabsorption of either TMAO or urea in the elasmobranch *Squalus acanthias* (20).

The origin of the TMAO that is found in the blood and body fluids of fish and other aquatic animals has been in dispute ever since the first demonstration of its occurrence in muscle of the dogfish, *Acanthias vulgaris,* by Suwa in 1909 (44). Since then it has been detected in relatively high concentrations in body fluids and tissues of many marine invertebrates, fishes and mammals, but there is little or none detectable in the freshwater species (9).

Table 2. *Reabsorption of trimethylamine oxide in 1.9 kg elasmobranch, Squalus, unaffected by the simultaneous presence of tetraethylammonium ion in doses exceeding those that competitively inhibited its secretion in the teleosts, Lophius (20)*

Time	Urine flow	Glomerular filtration rate	Plasma	Urine	TMAO Filtered	Excreted	Reabsorbed	Urea Reabsorbed	TEA
Hours	ml/kg/day	ml/kg/day	µmole/ml	µmole/ml	mmole/kg/day	mmole/kg/day	%	%	µmole/ml
0 – 2.2	30.5	79.4	89.2	11.1	7.08	0.34	95.2	95.0	–
2.2– 4.5	25.0	63.0	87.6	12.7	5.47	0.32	94.2	94.5	–
4.5– 7.4	17.2	53.1	85.7	8.2	4.65	0.15	96.8	96.2	–
			8.5 mg/kg TEA-Br intramuscularly at 7.4 hr						
7.4–10.0	17.9	46.0	83.7	10.6	3.88	0.19	95.2	95.0	2.5
			8.5 mg/kg TEA-Br intramuscularly at 10.0 hr						
10.0–12.7	16.8	48.7	81.9	6.7	3.96	0.11	97.2	95.0	2.5
12.7–15.2	31.4	52.8	80.1	2.2	4.23	0.07	98.3	97.8	0.9

When fish such as the salmon migrate between freshwater and marine environments the TMAO content of their tissues and body fluids fluctuates accordingly. The increase that occurs during the marine phase appears to be derived from dietary sources. When young salmon are raised on a TMAO-free diet and then kept on this diet as they are being transferred to increasing salinities the usual increase in TMAO content of tissues does not occur, even after 5 weeks in full strength seawater. Young control salmon that are instead fed on TMAO-containing food such as ground scallop muscle mixed with the usual diet rapidly accumulate the compound and within 3 weeks achieve TMAO levels characteristic of normal marine adults (3).

On the other hand, Bilinski (6) found that ^{14}C-labelled trimethylamine administered intraperitoneally in certain marine teleosts such as the lemon sole and the starry flounder could be recovered in the oxide. There was very little or no incorporation of ^{14}C in TMAO after the administration of several other free amines that on theoretical grounds might have been likely precursors of the oxide. It is of practical significance to note that intraperitoneal injections of the free amines gave better ^{14}C recoveries in the oxides than did similar doses administered intravenously or intramuscularly. Uptake from the latter sites in fishes is much faster than from the peritoneal cavity, and it appears that rapid excretion of the free base into the characteristically acid urine of these marine teleosts removed the precursors so fast that only an incomplete conversion of the administered dose into TMAO was possible by whatever tissues were involved. At any rate, the slow conversion of the methyl donors in marine teleosts as compared to the much brisker rates of incorporation in the marine lobster and other crustaceans, for example, raised doubt as to the relative importance of endogenous TMAO synthesis in fish tissues. In addition, the possibility of synthesis by microbial activity in the digestive tract of these flatfish was not ruled out as a likely source of this limited conversion.

An exogenous source seems likely for the very high tissue levels of TMAO found in marine elasmobranchs, despite the fact that plasma levels of TMAO in *Squalus*, for example, remain quite constant for as long as 40 days with the animals maintained without feeding under laboratory conditions. With the use of isotopically labelled precursors no incorporation was found of methyl ^{14}C-labelled trimethylamine or choline into TMAO, either *in vivo* or by liver slices *in vitro*. A test of possible incorporation of L-methionine(methyl-^{14}C) in the latter preparation was also negative. The assay procedures were sensitive enough to detect a rate of TMAO synthesis corresponding to the actual rate of excretion of this compound by the kidneys and gills of the dogfish (approximately 5 μmoles per kg X hr^{-1}) (24).

The essentially constant plasma levels observed in dogfish over a period of weeks of starvation is probably the result of two factors; the efficient reabsorption of TMAO by the renal tubules and the large pool of TMAO stored in muscle that can be drawn upon. The permeability of muscle tissue to TMAO is quite low. As shown in Table 3, the specific activity of ^{14}C-TMAO in muscle two days after intravenous administration of the labelled compound was only 5% of that in plasma. Even this small percentage could be accounted for in the

Table 3. *Slow rate of trimethylamine oxide entry into tissues of three 1.5 - 3.5 kg dogfish, Squalus, as disclosed by specific radioactivities of the oxide 2 and 20 days following the intravenous injection of 1.5 µCi ^{14}C-TMAO (24)*

Exp. No.	Days	Tissue	TMAO (µmoles/ml plasma or tissue water)	^{14}C-TMAO (cpm/ml plasma or tissue water)	Specific activity (cpm/µmole) TMAO	Percentage of plasma specific activity
1	2	Plasma	76	4727	62.2	–
		Muscle	225	745	3.3	5
		RBC	64	632	9.9	16
		Liver	87	3680	42.2	68
2	20	Plasma	45	730	16.2	–
		Muscle	154	764	5.0	31
		RBC	58	783	13.5	83
3	20	Plasma	78	1550	19.9	–
		Muscle	216	904	4.2	21
		RBC	85	1540	18.1	91

extracellular space of dogfish muscle (15%). Red blood cells after two days were 16% equilibrated with plasma, and liver, 68%. After 20 days the specific radioactivity of TMAO in muscle was only 20 - 30% that in plasma, when that in erythrocytes was 80 - 90%. It is apparent that muscle (and even erythrocytes) accumulate and presumably release TMAO very slowly. Thus, the slow release of TMAO from the large mass of muscle can probably account for the small amount of compound lost from plasma via gills and kidney during long periods without feeding. In the marine teleost (Table 1) an exogenous source for TMAO is also suggested by the spontaneous fall in TMAO excretion invariably noted under laboratory conditions during starvation.

Dogfish pups are born alive after a gestation period of almost two years in the cold waters of the Gulf of Maine. During this period in the mother's uterus growth and development occurs entirely at the expense of yolk without feeding and without functional communication with the mother's blood. It occurred to us that the absence of TMAO in fetal body tissues would point to an exogenous source if it then appeared later only after the newborn had begun free feeding. We found, however, that the concentration of TMAO in both yolk and in body fluids of embryonic pups was in the range of that normal for adult tissues, indicating, thereby, that retention of TMAO occurs very early in development and is a constant feature of the osmoregulatory operation in these marine elasmobranchs.

In other lower organisms N-oxidizing activity has been clearly demonstrated, and betaine, choline, and possibly lecithin and carnitine are precursors of amine oxides. The reduction of TMAO and the resultant liberation of volatile foul smelling trimethylamine in fish spoilage and in the feces of marine birds that feed on fish is probably due to bacterial N-oxide reduction. Despite the mystery regarding its fate and origin in most organisms, there can be little doubt that the relatively stable compound is cycled in and out of diverse organisms in the food web that comprises the marine faunal community, and that kidneys in at least two groups of organisms are not indifferent to it - marine teleosts that have tubular transport processes for actively secreting the compound, and marine elasmobranchs that actively reabsorb it. Its retention in body fluids and tissues of the latter plays a significant osmoregulatory role.

Several dozen other N-oxides have been reported as naturally occurring compounds in plants, animals and microorganisms, and in many cases the ability of animal tissues to synthesize the oxides from tertiary amine precursors has been clearly demonstrated. Nicotinamide, chlorpromazine, dimethylanaline, quanethidine, morphine, nicotinamide, nicotine and many other tertiary amine drugs have been shown to form N-oxides in animal tissues (4). There is very little known about the distribution, excretion or other translocation properties of many of the amine oxides, within the organism or in nature. Quantitative studies regarding rates of synthesis and subsequent translocation are difficult because of the striking differences in polarity and lipid solubility between the free amines and their corresponding oxides. The transformation of imipramine to its N-oxide, for example, involves a shift in pK_a from 8.0 to 4.7 (5). This results in movement of the oxide across biological membranes being quite independent of pH, in contrast to that of the amine precursor. The physiochemical differences between the free amine and the oxidized product thus strongly affect the respective distribution of the compounds within the organism and markedly facilitate rapid renal excretion of the oxides.

ORGANIC ANIONS

A wide variety of environmental agents originally with high lipid solubility and relatively low excretion rates are altered chemically

and ultimately fed into an organic acid transport system of wide occu
rence in excretory organs of both invertebrate and vertebrate animals
Tubular secretion of the foreign compound usually occurs only after
certain metabolic transformations and conjugation reactions have inte
vened that facilitate elimination of the compounds by enabling them t
associate with the transport carrier. Evidence for the existence of
such secretory systems in renal tubules was first obtained from Mar-
shell's classic studies on the excretion of phenol red (phenolsulfonphthalein), (30). This compound is exclusively a product of the
chemist's art, and for some time it appeared that nature's profligacy
had come up with a transport system in search of passengers. At the
same time in the microsomal fraction of the cells of the liver more
and more enzymes were turning up that seemed to be in search of substrates. Gradually it became clear that such cooperation between
renal secretory systems and hepatic detoxication mechanisms evolved
as adaptively significant devices that protected animals from the
toxic effects of naturally occurring compounds that were foreign
to them. Presently, with the rapidly increasing amounts of synthetic foreign compounds that find their way into the animal body as
food preservatives, medicines and environmental contaminants, it is
becoming apparent that these interlocking detoxication reactions and
excretory mechanisms are of critical importance in insuring well bein
or even survival.

The most widely studied compound in the organic transport system
is p-aminohippuric acid. It is used both to establish details concerning the kinetics of tubular secretory mechanisms, *in vitro* and

in vivo, and also clinically to measure effective renal blood flow and maximum tubular secretory capacity. The amino group in the *para* position is used merely to facilitate chemical determination; its presence does not appreciably affect renal handling of the parent compound, hippuric acid.

Interestingly, the first known conjugation mechanism was the biosynthesis of hippuric acid which was finally established by Keller in 1842 after many years of confusing the detoxication product with the precursor, benzoic acid, which is now very widely used as a food preservative. Benzoic acid is one of many aromatic carboxylic acids which are detoxicated by combination with amino acids, in this instance glycine, to form peptide conjugates. In the reaction (Fig. 2)

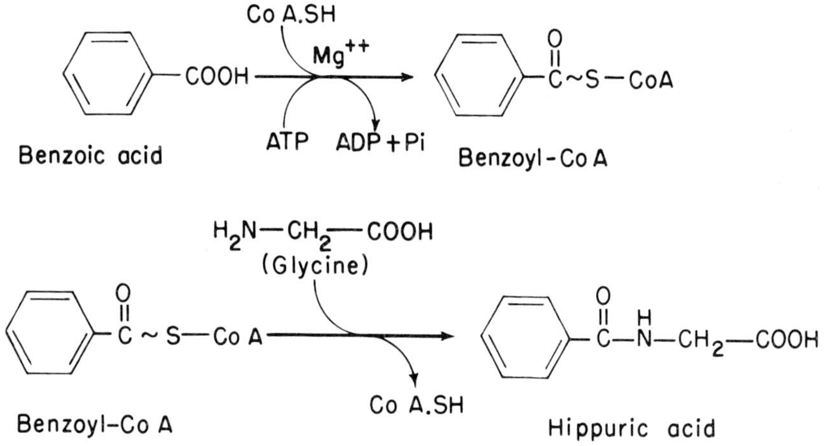

Fig. 2. Detoxication of benzoic acid by conjugation with glycine to form hippuric acid.

benzoyl-CoA is formed as an intermediate, and it is then conjugated with glycine and with certain other amino acids in various classes of animals. Peptide conjugation with amino acids occurs in both liver and kidney in most species; in dogs and chickens, however, it occurs only in the kidney. All terrestrial animals, fishes and crustaceans are capable of forming glycine conjugates with aromatic carboxylic acids. Birds and reptiles can also form ornithine conjugates, and spiders and myriapoda form conjugates with arginine and glutamine. Certain primates including man and the chimpanzee can also form glutamates from glutamine (32).

Other detoxication reactions that feed their nitrogenous products into the so-called "hippurate" renal transport mechanism are those involving conjugation reactions with glucuronic acid and with an activated form of sulfate (40, 41). Conjugation with glucuronic acid is the most important conjugation mechanism, and it occurs in all mammals and most vertebrates except fishes. Glucuronide conjugates result from formation of the coenzyme donor and then transfer of the glucuronyl moiety to phenols, alcohols, aromatic amines and carboxylic acids. Some very interesting complications sometimes arise as when one tries to identify the inhibitory effects of compounds such as 2,4-dinitrophenol which may interfere metabolically with the transport of anionic analogues in the organic acid transport system, by the parent compound's action in uncoupling aerobic phosphorylation and also competitively, by the glucuronide derivative's occupation of sites on the common carrier. The large important group of so-called "ethereal sulfates" are esters formed by the reaction of aromatic and aliphatic hydroxyl groups, and of certain amino groups, with activated sulfate. Other representative foreign nitrogenous compounds actively transported by the anionic "hippurate" system include penicillin, iodopyracet, probenecid and chlorothiazide. The anionic group is often a carboxyl group, but sulfones such as phenol red and chlorothiazide carrying partial negative charge are also briskly transported and they compete with the carboxylic acids for common carrier.

Because of technical complications involved in collecting bile, relatively little kinetic information is available concerning the biliary excretion of these compounds. Aside from preparing a compound for active tubular transport any hepatic detoxication reaction that increases polarity and diminishes the lipoid solubility of a compound will hasten its excretion from the organism in feces or by making it less readily reabsorbable by passive permeation from renal tubules subsequent to its removal from blood by glomerular filtration. Also, compounds that enter the bile duct in a non-polar form, or are rendered more lipid soluble by the actions of hydrolytic glucuronidases or sulfatases present in the intestinal flora, may well be repeatedly recycled by an enterohepatic circuit involving the gut, blood and liver (Fig. 3).

The organic anion transport systems of liver and kidney are not completely identical (49, 36). Some acids that are efficiently secreted by one organ are eliminated inefficiently or not at all by the transport process in the other. Glycocholic and taurocholic acids are the normally occurring biliary compounds in mammals, reptiles and birds. As Sperber pointed out (43), their high concentration in bile strongly suggests active transport from hepatic cells to bile. The glycocholic acids, like the hippuric acids, are glycine

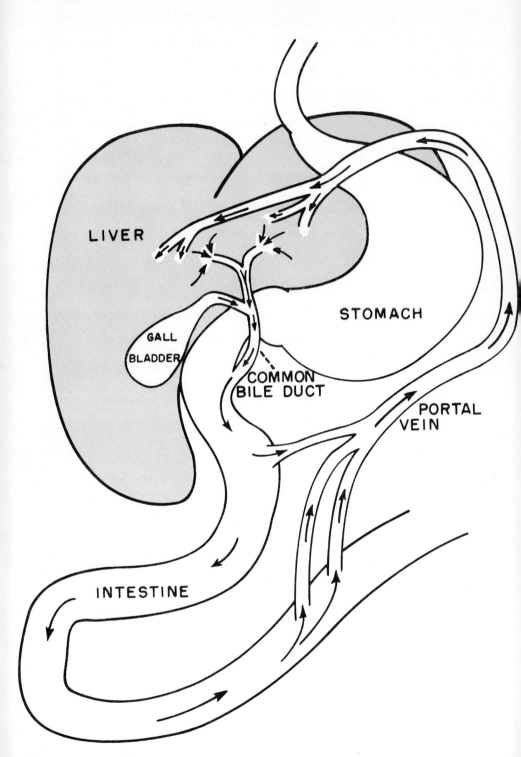

conjugates. The taurocholic acids have no counterpart in normal urine, but as sulphonic acids they are related to members of the phenolsulfonphthalein series which are actively transported by renal tubules. In the lower vertebrates bile contains the sulfuric esters of such "bile alcohols" as scymnol in elasmobranchs, ranol in the frog, and pentahydroxybufostane in the toad (25). Bilirubin, as well as the endogenous steroid alcohols and thyroxine, appear to be formed and excreted by the same pathways as the glucuronides of nitrogenous xenobiotics.

The renal handling of exogenous and other nitrogenous organic anions by the "hippurate" system is generally the same throughout the animal kingdom, but certain species differences do exist. The "brush-border" or proximal segment of the renal tubule is the site of active secretion in the vertebrates, and this portion of the nephron is a remarkably uniform and persistent structure from fish to man. In the invertebrates, however, two ontogenetically distinct tubules

---

Fig. 3 (p. 248). Enterohepatic cycle. Certain agents conjugated in the liver and excreted in bile are hydrolyzed in the gut by action of digestive enzymes or the intestinal flora and, if the resultant product is in a form favorable for intestinal reabsorption, it may then be carried via the hepatic portal circulation to the liver where it is again metabolized and excreted into bile as the reconjugated product, only to reenter the gut where it may again be hydrolyzed and recycled in its relatively non-polar state.

of entirely different embryological origins in various groups, the true nephridia and the coelomoducts, exhibit a remarkable case of functional convergence with respect to their tubular secretory functions (16). Transport of the organic acids in the system is dependent upon aerobic phosphorylation as the source of free energy, the temperature coefficient is characteristic of first order biological reactions, and transcellular movement against steep gradients of chemical activity is characterized by a maximal rate (T_m) and subject to competitive inhibition (Figs. 4 and 5).

Kinetic studies interpreted according to the Michaelis-Menten carrier hypothesis show that when the carrier is relatively unloaded the transport rate is proportional to the affinity of substrate for carrier, but when the carrier is near saturation the rate is inversely proportional to affinity. When the carrier is nearly loaded it is dissociation of the complex, rather than its formation, that determines the overall transfer rate of free substrate. When two transported substances are simultaneously competing with one another, and the carrier is far from being saturated by both substances, the substrate with higher affinity will have the faster transfer rate. At very high concentrations, however, the carrier will be saturated with the high affinity substrate, and the transfer of the substrate with lower affinity will be completely blocked. Effective competitors have high affinity for the carrier; they easily saturate it, but at the same time they are only slowly transported themselves (18, 19).

The main differences in the handling of organic anions by the

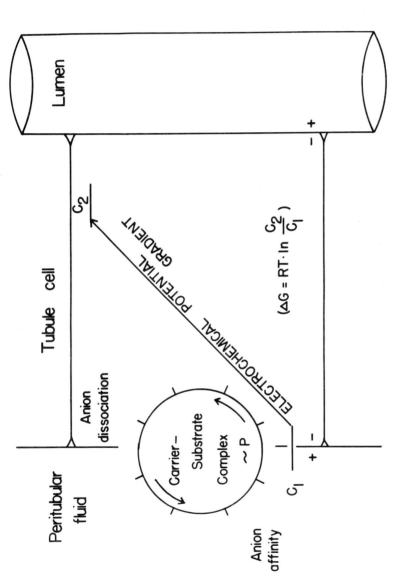

Fig. 4. Carrier-mediated active transport of organic anions against an electrochemical gradient is coupled to the hydrolysis of ATP, subject to competitive inhibition, and characterized by a maximal secretory capacity (T_m). (From Forster [18]).

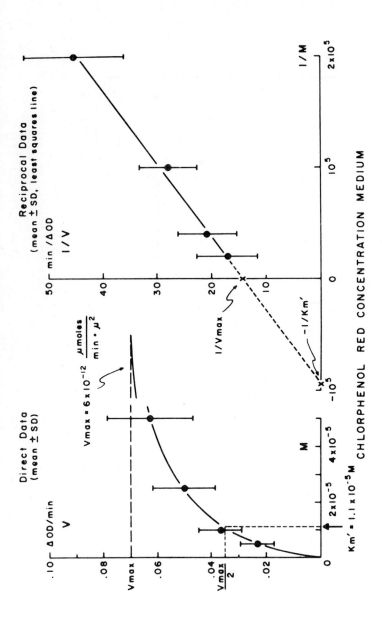

Fig. 5. Anionic dye transport into the lumen of isolated flounder tubules. Velocity (V) and concentration (M) data follow the pattern of classical saturation kinetics (Michaelis-Menten), and the reciprocal plot (Lineweaver-Burk analysis) provides numberical estimates of V_{max} and K_m'. (From Kinter [28]).

renal tubules of various vertebrates appears to be in the number of steps involved in transcellular movement, and in the substrate-carrier association and dissociation characteristics at each site. Direct observations made of movement across the cell from the vascular to the luminal side of *in vitro* preparations from such cold-blooded animals as the flounder clearly indicate active transport cell processes in membranes both at the serosal and mucosal sites (26, 18). In mammalian preparations, however, intracellular uptake on the vascular side of the cell clearly is an active process but there is no clearcut evidence that further accumulation occurs on the luminal side of proximal tubular cells (21, 46, 47) (Fig. 6).

The concept of bidirectional active transport originated with Kinter's observation on *Necturus* that some individuals showed net secretion of the foreign iodinated organic anion, Diodrast, and some showed net reabsorption (Fig. 7). Both processes were characterized by maximal transfer capacity and both were subject to competitive inhibition (27, 45). More recent evidence indicates the existence of such bidirectionally oriented pumps in other species, including mammals. For example, the reabsorptive flux of p-aminohippuric acid in dogs can be inhibited by probenecid (10), and secretory fluxes of urate (50) and taurocholate (51) have also been reported which oppose the over-all net active reabsorption of these compounds. Stop-flow experiments clearly indicate a bidirectional active transport of m-hydroxybenzoate in the proximal tubule of dogs (31), and of urate in the rabbit (2).

It appears that differences in the transcellular movement of a

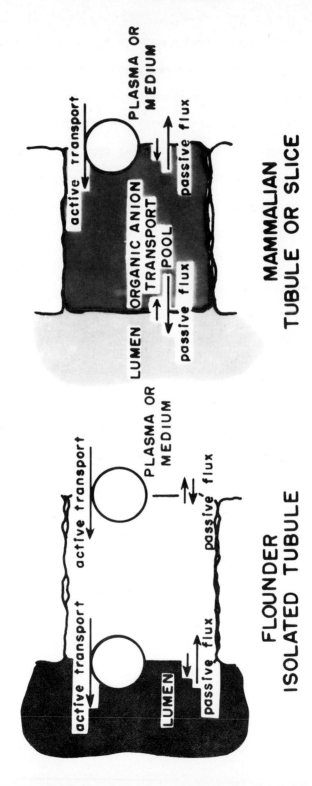

Fig. 6. Different models for secretion of organic acids in isolated flounder tubules and in mammalian *in vitro* preparations of proximal tubule cells.

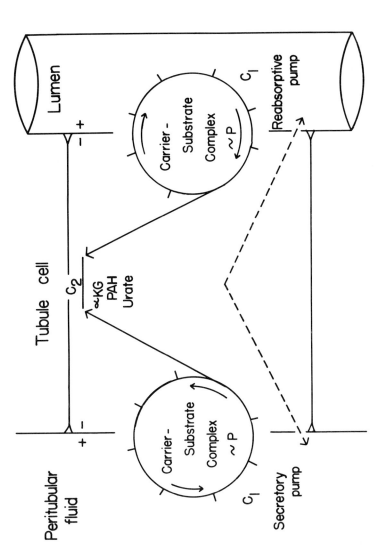

Fig. 7. Bidirectional active transport with pumps at the peritubular and luminal surfaces are oriented to move Krebs cycle intermediates, PAH and Diodrast into proximal tubule cells of *Necturus*. Evidence exists for the bidirectional movement of urate and certain other organic acids into proximal cells of mammalian species as well. (From Forster [18]).

specific compound among species can be due to the relative importance of the secretory and the reabsorptive components of the bidirectional transport process, and to the graded affinities that the respective carriers may have for the various competitors in the transport system.

It is quite unlikely that the primary function of an elaborate bidirectional mechanism such as this is merely to control excretion rates of exogenous organic acids. The direction inward of both of these transport processes suggests as a reasonable hypothesis that their real significance concerns their involvement in a system of intracellular homeostasis that serves mainly to control the delivery and to maintain levels of endogenous acidic metabolites or oxidative intermediates. Bidirectionally oriented systems such as this probably operate in many cells with two "faces", but they are revealed more clearly in the renal tubule because of the effective means that are available to evaluate transport processes here that operate separately at the basal and free surfaces of these cells. Evidence exists, for example, that the Krebs tricarboxylic cycle intermediate, α-ketoglutaric acid, which is normally reabsorbed may also enter the cell by secretion from the vascular side of proximal tubule cells. Probenecid blocks its secretion, and α-ketoglutarate inhibits the uptake of p-aminohippuric acid.

In the liver also, it has been suggested (38) that the uptake of α-ketoglutarate and other intermediates of the tricarboxylic acid cycle is normally via the anionic "hippurate" active transport system. Citrate is also normally reabsorbed by the proximal tubule, but experiments with labelled citrate show that it can move in both directio across the luminal surface of proximal cells, with reabsorption predominating. Succinate inhibits citrate reabsorption and interferes to some degree with its movement in both directions (12). Octanoate and other fatty acids inhibit the reabsorptive "hippurate" pump in *Necturus*, and competitively may convert net reabsorption of p-aminohippurate or Diodrast, in instances where it does occur, to net secretion. A circulating factor has been postulated whose action in this case would be stimulated by fatty acids, and it is viewed as determining the direction of net transport by selectively inhibiting the reabsorptive pump (45).

Whatever the specific mechanism might be, it appears that both the liver and kidney are prepared to handle foreign nitrogenous anions they have never "seen" before by retooling ancient biochemical reactions and transport processes that cooperate not only to eliminate these exogenous compounds efficiently but also first to alter them metabolically and then to convert them to appropriate conjugated derivatives before doing so.

ORGANIC BASES

All of the compounds secreted by the organic cation system in the renal tubules of vertebrates are nitrogen containing compounds. Some of the rather diverse members of this series have more than one positive charge, and the list includes quarternary ammonium compounds, amines, aliphatic and aromatic agents, and endogenous as well as exogenous compounds (Fig. 8). The organic base system is entirely independent of the organic anion active transport system as far as competition for common carrier is concerned, but there are many features

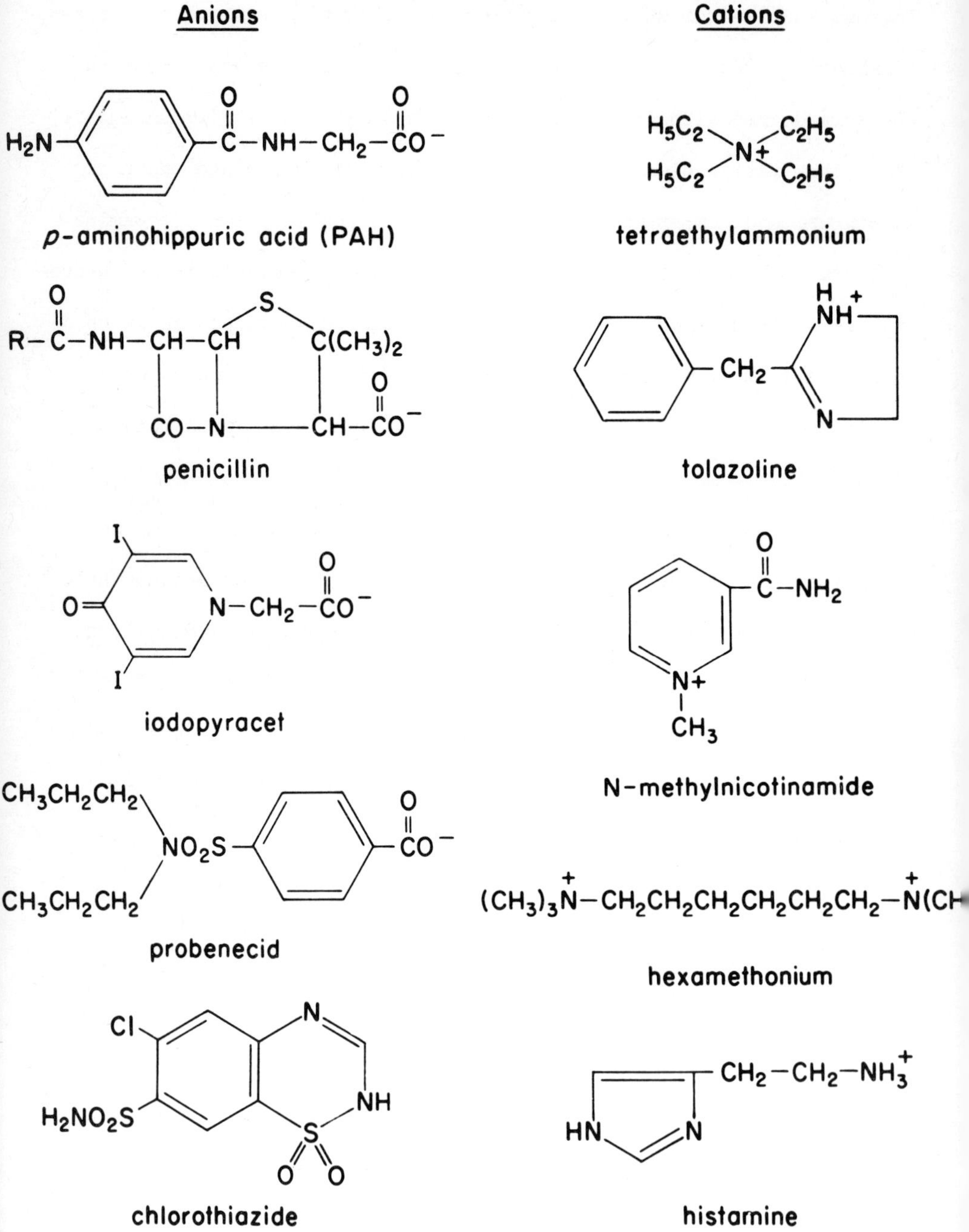

Fig. 8. Some organic nitrogenous acids and bases actively transported by renal tubules. (From Goldstein et al. [22]).

in common; both are located exclusively in the brush-border proximal segment of the renal tubule, secretion is characterized by a maximal rate (T_m) and, in some, passive reabsorption follows or accompanies active secretion. Michaelis-Menten kinetic treatment applies; both processes are energy dependent, stereospecific and subject to competitive inhibition (39, 35, 33).

Most of the exogenous and endogenous compounds in the base secreting system have powerful physiological effects on the organism (*e.g.*, choline, dopamine, histamine, serotonin, morphine, procaine, tetraethylammonium ion). The chicken renal portal preparation has been particularly useful in evaluating renal tubular handling of such agents because by this technique excretion can be measured *in vivo* using very small doses, and the tubular secretory process is invoked before the drug has an opportunity to exert a generalized systemic response. This clever application of a comparative preparation makes use of the renal venous blood supply which in birds and all the other lower vertebrates by-passes the glomeruli and is delivered exclusively to the renal tubules. Following separate ureteral catheterization a compound is introduced into a leg vein which supplies the capillary network surrounding tubules of the ipsilateral kidney exclusively. If the foreign agent undergoes tubular secretion, that fraction of the total amount injected will be excreted only by the ipsilateral kidney during this single passage. What remains of the injected dose will be delivered to both kidneys after completing the heart-lung-heart circuit. The difference between the amounts excreted by the ipsilateral and contralateral kidneys divided by the dose injected will represent the fraction removed by tubular secretion (40, 42).

There are other nitrogen containing agents such as the catecholamines that seem to be transported in part by the organic base system but, in addition, there appears to be another transport carrier specific for the catechol part of the molecule which is independent of both the acid and the base systems (34).

NONIONIC DIFFUSION

The doubly perfused frog kidney was the preparation that served to establish the first clear distinction between active transport and passive diffusion as tubular "secretory" processes. Chambers and Kempton (8) noted that phenol red (phenolsulfonphthalein) was transported by a pH-independent process whereas the elimination of neutral red (aminodimethylaminotoluaminozine hydrochloride) was strongly influenced by pH. Transfer rates of the latter appeared to be determined by the relatively high lipoid solubility of the nonionized moiety. Ammonium chloride and sodium bicarbonate were used to adjust the pH either of the tubular urine (via aortic perfusion), or of the peritubular blood (via perfusion of the renal portal vein). It was noted that the more acid the urine, the greater the amount of neutral red excreted, and that increases in alkalinity suppressed rates of dye elimination (Fig. 9). They then emphasized the passive nature of this kind of "diffusion and capture" by using an excised frog's bladder to show that neutral red accumulates within the bladder when pH is higher in the outside bath, and that runout of dye occurs when the external solution is more acid than bladder urine.

This kind of passive transfer across membranes where the movement of electrolyte down a gradient is determined by the degree of ionization and the relatively high lipoid solubility of the nonionized molecule is not restricted to liver and kidney. Absorption from the gut, gill or body surface may be influenced similarly, as well as is the distribution of such compounds between various tissues and body fluids within the organism. The pH of urine in marine fishes is normally two full pH units below the pH of plasma, and in mammals there is frequently a difference of 3 units. This means that potential concentration ratios of 100 or 1000, respectively, can be generated for weak organic bases having pK's in the range of plasma pH.

It must be pointed out, however, that it is an oversimplification to consider the uncharged moiety to be entirely lipid soluble, and also be view the cell membranes simply as bulk lipid. Also, the movement of a particular organic anion or cation across cell membranes may involve both active and passive processes. There are now many instances in both liver and kidney where it has been shown that a given compound may be secreted by an energy-dependent carrier-mediated system and also be subject to nonionic diffusion (43, 48). Anomolous movements of exogenous compounds across cells in gills, skin, gut, blood-brain barrier, liver and kidney result from the fact that all cell membranes are charged multiphasic systems having active and passive components, and that transcellular movement involves passage across at least two such complex heteogeneous units.

EXCRETION ACROSS GILLS

In aquatic organisms the excretion of nitrogenous and other

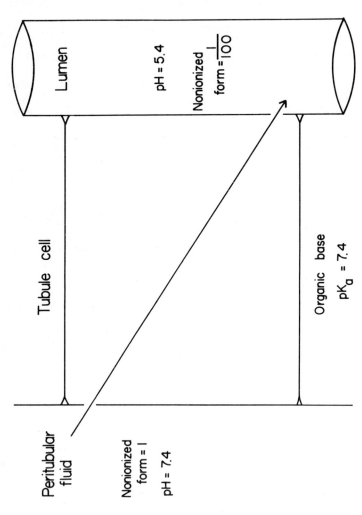

Fig. 9. Weak base depicted here is half ionized and half in the diffusible nonionized form in peritubular fluid with pH 7.4. In acid urine of pH 5.4, however, only about 1% will be in the diffusible form and a strong gradient is established in the direction of the lumen which results in rapid excretion of the base by passive diffusion in addition to glomerular filtration. (From Forster [18]).

foreign compounds may be greatly affected by the branchial circulation. In fish the entire cardiac output first traverses the gills before being distributed systemically, as with CO_2 and ammonia, and it is possible that certain exogenous compounds may be entirely cleared from blood by this means without biliary or renal excretory mechanisms being appreciably involved. Indeed, it has been suggested that microsomal detoxication mechanisms in the liver evolved mainly as adaptive devices for terrestrial living during evolution in much the same way that the conversion from ammoniotelism to ureo- or uricotelism occurred when the vertebrates abandoned their aquatic habita for land-dwelling. It was reasonably postulated that lipid-soluble organic compounds not readily excretable by terrestrial forms may not have been a problem to fish because they could be rapidly eliminated by diffusion through the lipoidal gill epithelial membranes into the effectively limitless external environment (7). This attractive hypothesis also considered the possibility of evolutionary regression during which such biliary enzymatic functions may have disappeared when genes were lost that were no longer needed to confer an advantage in natural selection. Aquatic turtles, for example, were cited as possible examples where such regression may have accompanied their migration from land back to the sea. Turtles adapted for life under water use their highly vascularized pharyngeal cavity as an accessory branchial structure to facilitate respiratory gas exchange, and this might obviate the earlier needs they had for detoxication reactions while air breathing on land.

Subsequent studies on the biogenesis of conjugation and detoxication products in a wide variety of fishes, however, have revealed many exceptions to this provocative speculation. Not only are aquatic forms capable of metabolizing a wide variety of foreign compounds but studies of the actual rates of excretion and half-lives in plasma of various drugs show also that gill excretion is relatively slow unless lipid solubility is very high (1).

Maren et al. (29) used a divided box preparation (Fig. 10) to measure the excretion of 8 foreign nitrogenous compounds with varying physicochemical characteristics and the relative contribution of branchial and renal processes to their elimination was determined. Table 4 presents the results of this study on the dogfish, *Squalus acanthias*. The findings conform fairly well with what might be considered a classic pattern of excretion for aquatic vertebrates: the lipid-insoluble ionized anion type (benzolamide) is actively secreted by renal tubules and is essentially impermeant at the gill. Weaker acids that are only slightly lipid-soluble and having pK's in the middle range (sulfanilamide and methazolamide) are neither secreted by the kidney nor rapidly eliminated at the gill. These agents are retained within the animal for long periods of time, and it seems that many foreign nitrogenous compounds are in this group. Another category consists of those that are essentially non-ionizable, either because they are non-polar in nature (*W 1100* and meprobamate), very weak proton-donators (ethoxzolamide) or proton-acceptors (antipyrine and tricain methane sulfonate or *MS 222*). With members of this group renal excretion is negligible, and their clearance by the kidney is always less than the glomerular filtration rate. In this group of weakly ionized compounds the rates of gill excretion are related to

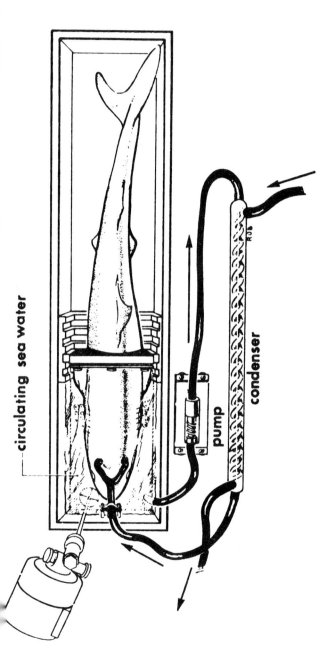

Fig. 10. Sea water in the head chamber is separated from that in the posterior chamber by a rubber collar around the dogfish. Water at 15° C is pumped through the condenser acting as coolant, and it is then introduced through the spiracles and across the gills. Not shown is a small aerator used in the head chamber where 1.5 to 2.0 l of circulating sea water is stirred by the motor. The rear compartment contains about 3 – 5 l of sea water. (From Maren et al. [29]).

Name and (n) [a]	Structure	pK_a	Partition $CHCl_3$ (pH 7·4 buffer)	Plasma $t^{\frac{1}{2}}$ free swim [b] (hr)	Urinary clearance Free [b] (ml/hr)	Urinary clearance Box [c] (ml/hr)	Gill clearance first hr (ml/hr)
Ethoxzolamide (4)	benzothiazole with SO_2NH_2 and C_2H_5O	8·1	27	1·3	3	3	84
Antipyrine (3)	phenyl pyrazolone with two CH_3	1·4	28	4	1	0·2	84
Tricain methane sulfonate (MS 222) (6)	benzene with $COOC_2H_5$, NH_2; CH_3SO_2H Salt	3·5	312	0·6	3	0·2	540

[a] (n) = number of experiments for each drug.
[b] Refers to the free swimming fish in the live car.
[c] Refers to fish in divided box, for first few hours.
[d] From (29).

Comparative gill and urinary excretion rates for selected foreign nitrogenous compounds with diverse physicochemical characteristics in the dogfish, Squalus acanthias[d]

Name and (n)[a]	Structure	pK_a	Partition $CHCl_3$ (pH 7·4 buffer)	Plasma $t^{\frac{1}{2}}$ free swim[b] (hr)	Urinary clearance Free[b] (ml/fish)	Urinary clearance Box[c] (ml/hr)	Gill clearance first hr (ml/hr)
Benzolamide (CL 11,366) (2)	[structure: phenyl-SO₂-N(H)-thiadiazole-SO₂NH₂]	3·2	0·0003	4	60	15	2
Sulfanilamide (2)	[structure: H₂N–C₆H₄–SO₂NH₂]	10·4	0·08	24	4	2	4
Methazolamide (1)	[structure: CH₃–C(=N–N(CH₃))–thiadiazoline–SO₂NH₂, C=O]	7·4	0·06	24	2	1	10
W 1100 (1)	NH₂OCOCH₂C(CH₃)(CH₃)CH₂OCONH₂	None	0·17	—	—	—	<30
Meprobamate (1)	NH₂OCOCH₂C(CH₃)(CH₂CH₂CH₃)CH₂OCONH₂	None	2·7	—	—	—	42

the chloroform-water partition coefficient; a 10^3-fold increase in lipoid solubility corresponded to a 10^2-fold increase in the gill excretion rate. Among these permeant compounds is the widely used anesthetic for aquatic organisms, "*MS 222*". Its gill clearance is almost one-half the cardiac output.

Current studies show that DDT and related organochlorine pesticides that have extremely high lipid-solubility, and are essentially insoluble in water, have rapid decay rates in plasma after intravenous administration to aquatic organisms but, paradoxically, they are eliminated practically not at all from the organism by either the branchial or renal route. The short half-life in plasma occurs here because the pesticide rapidly enters and becomes sequestered in intracellular lipid compartments and fatty tissues, mainly the liver, rather than being eliminated from the fish by diffusion from blood into the aqueous external milieu. In *Squalus acanthias*, for example, after 4 hr 65% of the dose was recovered from liver and the remaining 35% was distributed widely within the organism, mostly in muscle; in 48 hr 80% was in the liver and 15% of the dose in muscle, and then after 72 hr almost all of the DDT (96%) was more or less premanently situated in the large oily liver of this marine elasmobranch (T.H. Maren and D. Dverchik, personal communication). It is clear that foreign compounds need at least a certain degree of water solubility in order to enter the external aqueous environment besides the solubility in lipid requisite to entry of the cell membrane.

A NOTE ON RENAL HEMODYNAMICS

There is a wide difference between mammals and cold-blooded vertebrates with respect to the fraction of cardiac output delivered to the glomeruli, and this has a marked influence on the relative importance of roles that glomerular and tubular mechanisms play in the renal excretion of foreign compounds. In mammals more than 5% of the plasma cardiac output appears as glomerular filtrate under resting conditions (50% of the cardiac output goes to liver and kidneys). In man, for example, the entire plasma volume is filtered every 25 min, or 58 times a day. It is quite apparent here that excretion by glomerular filtration is far and away the most important route for the elimination of freely filterable foreign agents or their derivatives. Cardiac output in *Squalus acanthias* and in such marine teleost fish as the sculpin, *Myoxocephalus scorpius*, for example, average about 1,600 ml per kg per hr (23). Glomerular filtration rates in marine teleosts range from zero in the aglomerular forms to values that may go almost as high as in the elasmobranchs (15). Glomerular filtration rates for *Squalus* shown in Table 2 are typical. Here with a plasma cardiac output of 1,280 ml per kg per hr (20% hematocrit) the glomerular filtration rate is about 2 ml per kg per hr, so only about 0.15% of the cardiac output of plasma is filtered, as compared with about 5% in mammals. The fact that the fraction filtered is less than 3% of that in mammals emphasizes the relative importance of tubular secretion in these marine forms, and also the significance of extrarenal excretory processes. Despite the presence of a very active organic acid secretory system in such marine fish as the sculpin, *Myoxocephalus octodecimspinosus*, for example, approximately one-third as much phenol red can be accounted for in bile and gut as in the urine following intravenous administration of the dye (15). The fraction of cardiac output filtered is considerably higher in the freshwater vertebrates and in other inframammalian terrestrial forms, but in all of them the renal portal circulation that supplies the tubules exclusively comprises a large fraction of the total renal blood supply; and this venous supply becomes the predominant or sole source of blood to the kidney when the glomerular filtration rate may be sharply reduced in these lower vertebrates as in response to such well-documented environmental situations as cold, dehydration (15) and salt loading (13).

CONCLUSION

It seems that in this era with the rapidly increasing amount and variety of synthetic chemicals entering the air and water all of us who are concerned with the inter-relationships of animals and their environment will be functioning to some degree as pharmacologists. Knowledge of general principles that apply to the detoxication, distribution and elimination of foreign compounds will be essential, but comparative considerations, however, must also be kept in mind. Model based strictly on the behavior of drugs in mammalian systems are not entirely adequate to explain the handling of xenobiotics by invertebrates and by cold-blooded or aquatic vertebrates whose structural, functional and metabolic differences provide us with many interesting and significant variations on basic themes.

SUMMARY

Elimination of an exogenous agent is usually facilitated by detoxication processes or other metabolic interactions that speed its excretion by altering lipoid solubility, increasing ionization or by forming conjugates such as amino acid derivatives, glucuronides or ethereal sulfates which then are very efficiently excreted by specific active transport processes. Sometimes, as with trimethylamine oxide in sharks and other cartilaginous fishes, retention of the detoxicated product by an active renal tubular reabsorptive process serves a secondary role in osmoregulation. The comparative approach has proven to be of great value in physiology and pharmacology, not only in revealing roles that foreign nitrogenous compounds may play in special environmental situations, but also in providing effective models for the study of fundamental excretory mechanisms that are of significance generally. The elimination of amine oxides, organic acids and bases and other nitrogenous xenobiotics by renal, biliary and branchial systems are discussed, and excretory mechanisms in various species are analyzed in terms of filtration, "diffusion and capture," and carrier-mediated transport.

The excretion of foreign nitrogenous substances by inframammalian species will be emphasized here, especially in aquatic forms where a very close relationship exists between the external environment and the internal milieu. Many of the lower vertebrates serve as useful models in physiological and pharmacological experiments, and illustrations will be given where the comparative approach has been particularly fruitful in establishing concepts of general significance.

We will also examine how these organisms affect, as well as are affected by, the ever increasing quantity and diversity of exogenous chemical agents, or xenobiotics, that are presently accumulating in our marine, freshwater and estuarine environments.

The processes by which foreign nitrogenous compounds are eliminated are very closely related to metabolic detoxication mechanisms that modify the physical and chemical characteristics of these agents, directly serving thereby to protect the organism and also to promote the more rapid elimination of these compounds. However, especially with the vast numbers of bottom dwelling microorganisms and invertebrates, the chemical environment itself may be altered by the biota. For example, there are now well documented cases showing that water insoluble agents settle to the bottom where they accumulate in very high concentrations on the surfaces of detritus and other particulate matter. These compounds may then become highly concentrated in fatty tissues of filter feeding organisms and eventually reenter the external environment in a drastically altered form following metabolic transformations.

Before being excreted by specific renal or biliary processes, foreign nitrogenous compounds frequently undergo chemical reactions, usually in the liver, that introduce groups or expose functional focal points for secondary conjugation reactions. Metabolic modifications such as these increase glomerular filtration through aqueous pores, diminish reabsorptive permeation in the renal tubules, facilitate transport by specific carrier mediated processes, and prevent enterohepatic recycling. Transcellular diffusion or active transport processes, as across branchial, hepatic or renal epithelial cells, are multiphasic systems involving several biological membranes, each of which is composed of complex heterogeneous units. Nonelectrolytes move across such interfaces by simple diffusion in accordance generally with their lipid solubility. The diffusion rate of an electrolyte, however, is affected by the dissociation constant (K_a), the pH of the media on both sides of the barrier which fixes the respective concentrations of the ionized and molecular forms of the dissociated electrolyte, and by the lipid solubility of the nonionized molecular form of the compound. The binding characteristics with plasma proteins may also markedly affect the biliary and renal excretory rates of blood constituents, particularly in glomerular filtration or other passive transport processes where the availability for excretion is directly dependent upon concentration of the compound in plasma water. These reactions that affect the absorption, distribution and elimination of foreign compounds are effectively and comprehensively presented in several recent general publications (22, 32).

REFERENCES

1. Adamson, R.H. *Fed. Proc.* *26*:1047, 1967.
2. Beechwood, E.C., W.O. Berndt and G.H. Mudge. *Am. J. Physiol.* *207*:1265, 1964.
3. Benoit, G.J. and E.R. Norris. *J. Biol. Chem.* *158*:439, 1945.
4. Bickel, M.H. *Pharmacol. Rev.* *21*:325, 1969.
5. Bickle, M.H. and H.J. Weder. *J. Pharm. Pharmacol.* *21*:160, 1969.

6. Bilinski, E. *J. Fisheries Res. Board Can. 21*:769, 1964.
7. Brodie, B.B. and R.P. Maickel. In: *Proceedings of the First International Pharmacology Meeting*, edited by B.B. Brodie and E.G. Erdos. New York: Macmillan, 1962. Vol. 6, p. 299.
8. Chambers, R. and R.T. Kempton. *J. Cell. Com. Physiol. 10*:199, 1937.
9. Cheymol, J., P. Chabrier, S. Dechezlepretre and F. Bourillet. *Biol. Méd. (Paris) 56*:124, 1967.
10. Cho, K.C. and E.J. Cafruny. *Pharmacologist 9*:208, 1967.
11. Cohen, J.J., M.A. Krupp and C.A. Chidsey, III. *Am. J. Physiol. 194*:229, 1958.
12. Cohen, R.D. and R.E.S. Prout. *Clin. Sci. 28*:487, 1965.
13. Dantzler, W.H. and B. Schmidt-Nielsen. *Am. J. Physiol. 210*:198, 1966.
14. Farah, A., M. Frazer and E. Porter. *J. Pharmacol. Exp. Therap. 126*:202, 1959.
15. Forster, R.P. *J. Cell. Comp. Physiol. 42*:487, 1953.
16. Forster, R.P. In: *The Cell*, edited by J. Brachet and A.E. Mirsk New York: Academic Press, 1961. Vol. 5, p. 89.
17. Forster, R.P. In: *Sharks, Skates and Rays*. Baltimore, Maryland: The Johns Hopkins Press, 1967. p. 187.
18. Forster, R.P. *Fed. Proc. 26*:1008, 1967.
19. Forster, R.P. *Bull. Mt. Desert Is. Biol. Lab. 8*:29, 1968.
20. Forster, R.P., F. Berglund and B.R. Rennick. *J. Gen. Physiol. 42*:319, 1958.
21. Forster, R.P. and J.H. Copenhaver, Jr. *Am. J. Physiol. 186*:167, 1956.

22. Goldstein, A., L. Aronow, S.M. Kalman. *Principles of Drug Action*. New York: Harper & Row, 1968. p. 199.
23. Goldstein, L., R.P. Forster and G.M. Fanelli, Jr. *Comp. Biochem. Physiol. 12*:489, 1964.
24. Goldstein, L., S.C. Hartman and R.P. Forster. *Comp. Biochem. Physiol. 21*:719, 1967.
25. Haslewood, G.A.D. *Physiol. Rev. 35*:178, 1955.
26. Hong, S.K. and R.P. Forster. *J. Cell. Comp. Physiol. 54*:237, 1959.
27. Kinter, W.B. *Am. J. Physiol. 196*:1141, 1959.
28. Kinter, W.B. *Am. J. Physiol. 211*:1152, 1966.
29. Maren, T.H., R. Embry and L.E. Broder. *Comp. Biochem. Physiol. 26*:853, 1968.
30. Marshall, E.K., Jr. and J.L. Vickers. *Bull. Johns Hopkins Hosp. 34*:1, 1923.
31. May, D.G. and I.M. Weiner. *Am. J. Physiol. 218*:430, 1970.
32. Parke, D.V. *The Biochemistry of Foreign Compounds*. Oxford: Pergamon Press, 1968. p. 130.
33. Peters, L. *Pharmacol. Rev. 12*:1, 1960.
34. Quebbemann, A. and B.R. Rennick. *J. Pharmacol. Exper. Therap. 166*:52, 1969.
35. Rennick, B.R., G.K. Moe, R.H. Lyons, S.W. Hoobler and R. Neligh. *J. Pharmacol. Exp. Therap. 91*:210, 1947.
36. Schanker, L.S. *Pharmacol. Rev. 14*:501, 1962.
37. Schmidt-Nielsen, B. and R.P. Forster. *J. Cell. Comp. Physiol. 44*:233, 1954.

38. Sellick, B.H. and J.J. Cohen. *Am. J. Physiol. 208*:24, 1965.
39. Sperber, I. *Proc. 17th Internat. Physiol. Congress.* Oxford, 1947. p. 217.
40. Sperber, I. *Ann. Roy. Agric. Coll. Sweden 15*:317, 1948.
41. Sperber, I. *Scand. J. Clin. Lab. Invest. 1*:345, 1949.
42. Sperber, I. *Arch. Int. Pharmacodyn. 97*:221, 1954.
43. Sperber, I. *Pharmacol. Rev. 11*:109, 1959.
44. Suwa, A. *Arch. ges Physiol. 128*:421, 1909.
45. Tanner, G.A. and W.B. Kinter. *Am. J. Physiol. 210*:221, 1966.
46. Tune, B.M., M.B. Burg and C.S. Patlak. *Am. J. Physiol. 217*:1057, 1969.
47. Watrous, W.M., D.G. May and J.M. Fujimoto. *J. Pharmacol. Exper. Therap. 172*:224, 1970.
48. Weiner, I.M. and G.H. Mudge. *Am. J. Med. 36*:743, 1964.
49. Woo, T.H. and S.K. Hong. *Am. J. Physiol. 204*:776, 1963.
50. Zins, G.R. and I.M. Weiner. *Am. J. Physiol. 215*:411, 1968.
51. Zins, G.R. and I.M. Weiner. *Am. J. Physiol. 215*:840, 1968.

CONTROL OF RENAL AMMONIA METABOLISM[1]

Robert F. Pitts

Department of Physiology
Cornell University Medical College
New York, New York 10021

The excretion of ammonia by mammals including man is primarily concerned with urinary excretion of non-volatile acid, not with elimination of waste nitrogen. Accordingly rate of excretion of ammonia is large in acidosis, negligible in alkalosis. Since urinary ammonia is formed within renal tubular cells from precursors extracted from blood perfusing the kidney, some change in composition of the urine and/or some change in constitution of tubular cells must control production and excretion of ammonia. Fig. 1 illustrates these points (17).

These experiments were performed on one normal unanesthetized dog in two states of acid-base balance; in those described by the lower line, the dog was in normal acid-base balance; in those described by the upper line, the dog was in chronic metabolic acidosis.

---

[1]The portion of this work which originated in the author's laboratory has been generously supported by the National Heart Institute of the National Institutes of Health and by the Life Insurance Medical Research Fund.

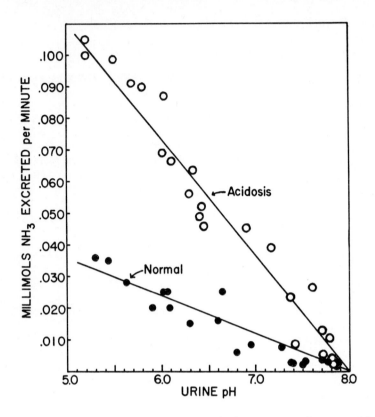

Fig. 1. The relation in the dog between ammonia excretion and urine pH in normal acid-base balance and in chronic metabolic acidosis. (From [17].)

At the beginning of each of the two series of experiments, the urines were comparably acid, pH 5.5 or below. Sodium bicarbonate was infused intravenously at rates required to correct the acidosis and gradually to alkalinize the urine. Urine pH increased from its initial low value to 8.0. Rate of ammonia excretion was inversely related to urine pH in both instances. However, at any given urine pH, ammonia excretion was some three times greater in acidosis than in normal acid-base balance. Obviously no less than two factors must control excretion of ammonia: first, urine pH; second, some adaptive change in the kidney induced by acidosis which increases production of ammonia. The relation of ammonia excretion and urine pH depends on the mechanism of its transport across the luminal membranes of tubular cells. This has been shown to be one of passive non-ionic diffusion down a pH gradient (6, 13). The adaptive increase in production of ammonia in acidosis concerns us here.

This adaptive increase in production and excretion of ammonia was first ascribed by Davies and Yudkin (5) to an adaptive increase in the renal content of the enzyme glutaminase. This inference was based on the now well established fact that the major precursor of urinary ammonia is glutamine. A number of investigators have confirmed a parallelism of enzyme content and ammonia excretion in the rat and guinea pig kidney.

This presumed causal relationship between an adaptive increase in renal glutaminase and ammonia excretion in acidosis became less convincing when it was shown by Rector and Orloff (20), Pollack *et al.* (19) and others that it did not occur in the dog. The thesis was finally demolished by Goldstein (7) who showed that the administration of Actinomycin D prevented enzyme adaptation without in any way altering the adaptive increase in ammonia excretion in acidosis. Obviously an increase in the glutaminase content of renal tubular cells is not obligatorily related to increased production and excretion of ammonia. However this does not mean that a fixed amount of enzyme may not increase its activity under conditions which exist *in vivo* in the acidotic kidney and thus actually produce more ammonia.

The production of ammonia by the kidney is not substrate limited in the usual sense, $i.e.$, the concentration of glutamine in the plasma entering the kidney, the rate of renal plasma flow and the rate of glomerular filtration are not different in chronic metabolic acidosis, normal acid-base balance, and metabolic alkalosis. However the utilization of the glutamine presented to the kidney does differ in these diverse acid-base states. The experiments of Pilkington et al. (15), described in Fig. 2A were performed on dogs to characterize the

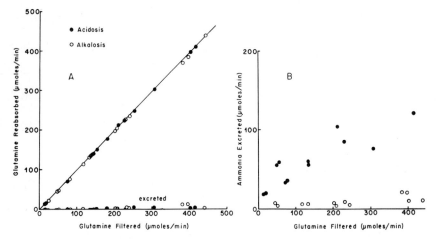

Fig. 2A. Reabsorbed and excreted glutamine as a function of filtered load in the dog in acidosis and alkalosis.

2B. Excretion of ammonia as a function of filtered load of glutamine in the dog in acidosis and alkalosis. (From [15].)

reabsorption of glutamine from the glomerular filtrate. The quantity of glutamine filtered is plotted on the abscissa. It was increased from the normal range of 25 to 30 µmoles/min to around 500 by the intravenous infusion of glutamine. The amount of glutamine reabsorbed was directly proportional to the amount filtered over the entire range in both acidosis and alkalosis. Essentially all was reabsorbed; none was excreted. Thus all of the glutamine which entered the glomerular filtrate crossed the luminal membranes and entered tubular cells independent of acid-base state.

What happens to this glutamine once it enters tubular cells obviously differs in acidosis and alkalosis. In Fig. 2B are shown data from these same experiments in which excretion of ammonia is related to glutamine filtered and reabsorbed. In alkalosis, once glutamine entered the cell it was transported across the cell relatively unchanged into peritubular blood. In acidosis, it was more or less metabolized to liberate ammonia. Thus some intracellular change in acidosis must account for increased tubular production of ammonia.

Are the amounts of glutamine which enter tubular cells from the glomerular filtrate sufficient to account for the ammonia produced by the kidney? In normal acid-base balance, the answer is yes; in acidosis, it is no. In acidosis, glutamine also enters tubular cells from peritubular blood and is metabolized to ammonia. Table 1 demonstrates in 6 experiments in dogs in normal acid-base balance, that the quantity of glutamine filtered was equal to 16.4 µmoles/min (15). Counting 2 nitrogens per molecule as potential sources of ammonia, the amount of glutamine reabsorbed from the glomerular filtrate exceeded the amount of ammonia produced. All of the glutamine reabsorbed was not utilized for production of ammonia, some was returned unaltered to renal venous blood, for

Table 1. *The relation between entry of glutamine into renal tubular cells from filtrate and from peritubular blood and ammonia produced by those cells in normal acid-base balance and in chronic metabolic acidosis.*

Condition	Glutamine (µM/ml)	GFR (ml/min)	Reabsorbed	Peritubular transport	Extracted	NH$_4^+$ produced
			(µmoles/min)			
Normal Acid-base Balance (6 exp.)	0.417	39.7	16.4	0	6.6	18.1
Chronic Metabolic Acidosis (13 exp.)	0.493	37.5	17.8	11.8	29.6	55.1

From [15].

only 6.6 µmoles were extracted by the kidney. These 6.6 µmoles could account for 13.2 µmoles of ammonia. Actually 18.1 µmoles of ammonia were produced so other precursors of ammonia in small amounts must be invoked to explain the ammonia produced.

In chronic metabolic acidosis 17.8 µmoles of glutamine per min entered tubular cells from the glomerular filtrate. An additional 11.8 µmoles per min were removed from peritubular capillary blood, for total extraction amounted to 29.6 µmoles/min. The sum of the two nitrogens of glutamine which entered tubular cells from filtrate and from peritubular blood, namely 59.2 µmoles/min, exceeded slightly the ammonia produced, 55.1 µmoles/min.

Kamin and Handler (11) and Orloff and Berliner (14) have pointed out that the moment to moment production of ammonia within tubular cells would relate best to its removal in urine and renal venous blood if it were produced by a reversible enzymatic reaction. The glutamate dehydrogenase reaction is one such possible reversible reaction which oxidatively deaminates glutamate and reductively aminates α-ketoglutarate. This reaction probably occurs in all mammalian kidneys. The glutamine synthetase reaction and the glutaminase-I reaction constitute an operationally reversible system in the rat, sheep, rabbit and guinea pig. Only the glutaminase-I reaction is operative in the dog and cat.

Fig. 3 illustrates the amination of ^{14}C-α-ketoglutarate *in vivo*

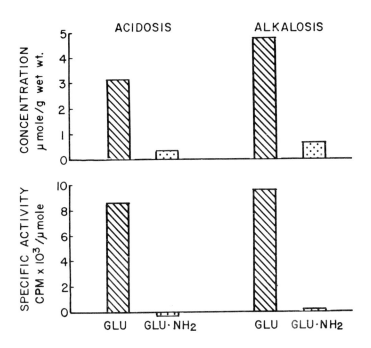

Fig. 3. Concentrations and specific activities of glutamate and glutamine in kidneys of dogs in acidosis and alkalosis

by the kidney of the dog to form ^{14}C-glutamate. It further illustrates the failure of synthesis of ^{14}C-glutamine from the labelled glutamate (12). These experiments were performed on dogs in acidosis (shown on the left) and in alkalosis (shown on the right). From 40 to 50 µCi of ^{14}C-α-ketoglutarate were infused into one renal artery at a uniform rate over a period of 40 min. At the end of this time the infused kidney was rapidly removed and homogenized in picric acid. A filtrate was prepared for flow scintillation analysis of ^{14}C activity and for automatic amino acid analysis by a system in tandem which permitted direct calculation of specific activity.

In the upper panels, concentrations of glutamate and glutamine in µmoles per g wet weight of kidney are shown. In the lower panels, specific activities in thousands of counts per min per µmole are shown. In the acidotic kidney, the concentration of glutamate averaged 3.15 µmoles per g and that of glutamine, 0.344 µmoles per g. In the alkalotic kidney, glutamate increased to 4.80 µmoles per g and glutamine increased to 0.631 µmoles per g.

In the lower panels the specific activity of glutamate was 8,265 cpm per µmole in acidosis and 9,565 in alkalosis. Thus ^{14}C-α-ketoglutarate was aminated to glutamate in both acidosis and alkalosis and to approximately the same extent. However no conversion of ^{14}C-glutamate to ^{14}C-glutamine occurred. This is consonant with the view that the kidneys of the dog and cat contain no glutamine synthetase.

(Fig. 3 [p. 283]) following 40 min infusion of ^{14}C-α-ketoglutarate into one renal artery. (Adapted from [12].)

In contrast the kidney of the rat contains the glutamine synthetase system. As shown in the upper panel of Fig. 4, the concentrations

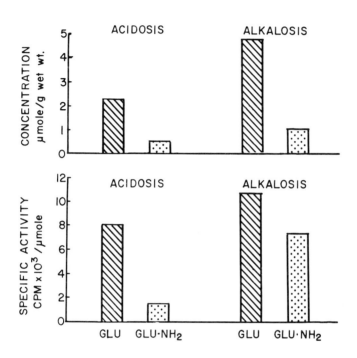

Fig. 4. Concentrations and specific activities of glutamate and glutamine in kidneys or rats in acidosis and alkalosis following 40 min infusion of ^{14}C-α-ketoglutarate into one renal artery. (Adapted from [12].)

of glutamate and glutamine in the kidney of the rat are roughly comparable to those in the kidney of the dog. When ^{14}C-α-ketoglutarate was infused for 40 min into one renal artery of the rat, the glutamate of the kidney was highly labelled exactly as in the dog. However the label was further incorporated into glutamine to a specific activity of 20% of that of glutamate in acidosis and to 70% of that of glutamate in alkalosis. These data suggest that in alkalosis the synthesis of glutamine constitutes a means of disposing of ammonia which is alternate to excretion. This could then constitute a system of control of ammonia excretion in the kidney of the rat, guinea pig, sheep and rabbit which does not exist in the cat and dog.

This view is borne out by studies of Damian and Pitts (4) in the rat on the rates of glutaminase-I and glutamine synthetase reactions *in vivo* in acidosis and alkalosis. Damian exposed the right kidney of a living rat and threaded a 31 guage needle across the aorta several millimeters into the right renal artery. He then pulse labelled the kidney by the rapid injection of 0.05 ml of saline containing from 0.5 to 1.5 µCi of ^{14}C-glutamine or ^{14}C-α-ketoglutarate. After 5 to 30 s the kidney was snatched out and thrown into liquid nitrogen. It was weighed, homogenized in picric acid and prepared for ^{14}C and amino acid analysis as in the experiments just described.

The thesis is that if one measures the concentration of glutamine and glutamate and the rate of transfer of counts from ^{14}C-glutamine to glutamate, one can calculate the rate of the glutaminase-I reaction. In other experiments the reverse transfer of counts from ^{14}C-labelled glutamate to glutamine permits calculation of the rate of glutamine synthesis. Actually α-ketoglutarate was administered in these experiments for its extraction is more complete than that of glutamate and it is essentially instantaneously aminated to glutamate.

In normal acid-base balance, the rates of the two reactions are roughly comparable: glutaminase-I, 22.5 µmoles/g/hr; glutamine synthetase 24.5 μmoles/g/hr. This fact can be correlated with a low rate of excretion of ammonia, *i.e.*, essentially all of the ammonia produced in the glutaminase-I reaction is utilized in the synthesis of glutamine.

In chronic metabolic acidosis, the rate of the glutaminase-I reaction exceeds the rate of glutamine synthesis. The glutaminase-I reaction is greatly increased to 83.2 μmoles/g/hr; glutamine synthesis is markedly reduced to 3.6 μmoles/g/hr. As a consequence, large amounts of ammonia are produced and excreted in the urine. Little is directed to synthesis of glutamine.

In chronic metabolic alkalosis, the reverse is true. The rate of synthesis of glutamine (26.5 μmoles/g/hr) exceeds the glutaminase-I reaction (7.6 μmoles/g/hr) and accordingly the excretion of ammonia is negligible. I hold no brief for the precision of our measured enzyme rates. Indeed these experiments constitute the first attempt to measure them *in vivo*. However rates so measured are more meaningful in some respects than rates measured *in vitro* on homogenates with gross excesses of substrates and cofactors. Indeed such *in vitro* rates on homogenates exceed those measured *in vivo* by 2 to 3 orders of magnitude. Furthermore *in vivo* rates at least approximate rates of excretion of ammonia and rates of addition of glutamine to renal venous blood in the rat. However the significant point I wish to make is that the glutaminase-I and glutamine synthetase reactions constitute an operationally reversible system sensitive to acid-base state. As such it provides a means of control of net renal production of ammonia. However I hasten to point out that although true in the rat and possibly in the sheep, rabbit and guinea pig, it is definitely not true in the dog and cat, which cannot synthesize glutamine. Whether the kidney of man synthesizes glutamine is unknown.

The three products of the glutaminase and glutamic dehydrogenase reactions, ammonia, glutamate and α-ketoglutarate, have all been implicated as controlling the glutaminase-I reaction through product inhibition. Since the evidence for a role for glutamate is the major thrust of the work of the next paper (see p. 297), I shall not consider it further. Rather I shall turn to ammonia.

In a series of experiments total renal production of ammonia, namely the sum of the ammonia excreted in the urine and added to renal venous blood, was measured in acidotic dogs before and during intravenous infusion of ammonium chloride (16). The control values are shown as the left hand points of the pairs connected by lines in Fig. 5. The values obtained during the infusion of ammonium chloride are shown as the right hand points. In each instance increased cellular pNH_3, resulting from the infusion of NH_4Cl, decreased cellular production of ammonia.

According to Goldstein and Schooler (9) the K_i for 50% inhibition of glutaminase by ammonia *in vitro* is some two orders of magnitude greater than pNH_3 values observed *in vivo*. Whether this means that the enzyme is more sensitive to inhibition by ammonia *in vivo* than *in vitro* or that the mechanism of inhibition is totally different and non-competitive is uncertain.

The isolated kidney of the acidotic rat can be perfused for two hours or more *in vitro* with Krebs-Henseleit solution containing 4.5% fraction V bovine serum albumin, to maintain proper oncotic relationships, and a single amino acid in 5 mM concentration, as substrate.

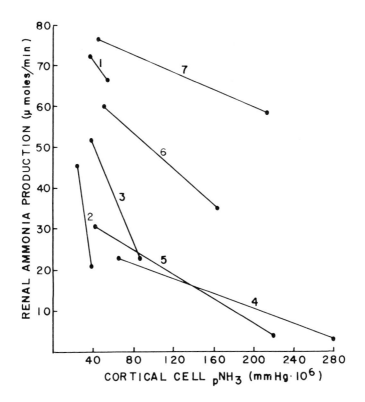

Fig. 5. The effect of increasing renal tubular cell pNH_3 on production of ammonia by the kidney of the dog. (From [16].)

During this interval the kidney remains metabolically viable in the sense that it utilizes the amino acid and adds to the perfusion medium, at rates which are linear functions of time, ammonia, glutamate, glutamine and glucose.

As shown in Table 2, rates of utilization of aspartate and rates of production of ammonia, glutamate, glutamine and glucose were measured in three experiments in acidotic rat kidneys in which only aspartate was present in the perfusion medium and again in three

Table 2. *The effect of addition of α-ketoglutarate on production of ammonia, glutamate, glutamine and glucose from aspartate by the perfused rat kidney*

NH_4^+	Glutamate	Glutamine	Glucose	Aspartate	α-Ketoglutarate
(μmoles/g dry wt X hr ± S.E.)					
Aspartate (5 mM) (3 exp.)					
390	97.0	15.5	76.0	- 476	0
± 37.9	± 39.0	± 8.0	± 10.0	± 57.4	
Aspartate (5 mM) + α-Ketoglutarate (5 mM) (3 exp.)					
132	208	83.3	254	- 515	- 686
± 45.0	± 54.3	± 13.3	± 38.4	± 74.1	± 34.3

From [18].

other experiments in which both aspartate and α-ketoglutarate were present (18). The production of ammonia decreased from a mean of 390 μmoles per g per hr to 130 μmoles per g per hr, namely to one-third, on the addition of α-ketoglutarate. The production of glutamate and glutamine increased roughly in proportion to the decreased production of ammonia. Thus the utilization of aspartate did not change significantly on addition of α-ketoglutarate; rather nitrogen previously added to the medium as ammonia was now added as glutamate and glutamine. This is not peculiar to aspartate alone; it is also true of kidneys perfused with leucine, glycine, alanine and no doubt other amino acids as well.

Table 3 demonstrates that 5 kidneys, perfused in the absence of all substrate, produced on an average 49.8 μmoles of ammonia per

Table 3. *The effect of addition of α-ketoglutarate on production of metabolites by the rat kidney perfused in the absence of all substrate and in the presence of ammonium bicarbonate*

	NH_4^+	Glutamate	Glutamine	Glucose	α-Ketoglutarate
	(μmoles/g dry wt. X hr ± S.E.)				
No substrate (5 exp.)	49.8 ± 3.2	0	0	6.4 ± 1.3	0
α-Ketoglutarate (5 mM) (3 exp.)	5.6 ± 1.1	13.7 ± 3.1	25.0 ± 8.1	240 ± 16.2	- 559 ± 49.2
α-Ketoglutarate (5 mM) NH_4HCO_3 (2 mM) (4 exp.)	- 138 ± 14.0	67.7 ± 8.1	51.9 ± 4.4	392 ± 43.0	- 1007 ± 50.2

From [18].

g per hr from endogenous nitrogen sources (18). No glutamate or glutamine was formed. When 5 mM α-ketoglutarate was incorporated in the medium as the sole substrate, the production of ammonia decreased to 5.6 μmoles per g per hr. Instead the production of glutamate and glutamine increased to make up the deficit in production of ammonia. When both 5.0 mM α-ketoglutarate and 2.0 mM ammonium bicarbonate were incorporated in the medium, ammonia was utilized at a rate of 138 μmoles per g per hr. Glutamate, glutamine and glucose were synthesized at greatly accelerated rates. If one adds the μmoles of glutamate and glutamine produced plus twice the μmoles of glucose produced, the sum is equal to 90% of the α-ketoglutarate utilized.

These experiments emphasize the fact that the synthesis of glutamate and glutamine provide a means of disposing of renal ammonia alternate to excretion. They also demonstrate that the rate of synthesis of these two amino acids is dependent on the availability of α-ketoglutarate. Goldstein (8), Alleyene (1) and Balagura-Baruch (2) have shown that the concentration of α-ketoglutarate in kidneys of acidotic dogs and rats is very low and in the kidneys of alkalotic dogs and rats is appreciably higher. True enough, the availability of α-ketoglutarate under *in vivo* conditions never approaches that attained in the kidney perfused with 5.0 mM α-ketoglutarate. However availability of α-ketoglutarate may be one of several factors controlling ammonia excretion.

Although these experiments do not demonstrate repression of glutaminase-I by α-ketoglutarate those of Balagura (2) on the dog show a major effect. She demonstrated that the intravenous infusion of α-ketoglutarate suppressed renal production of ammonia in proportion of plasma concentration. The increase in kidney concentration of glutamate and the increase of glutamate added to renal venous blood were together equivalent to only one-quarter of the reduction in ammonia production. Thus α-ketoglutarate must suppress activity of glutaminase as well as control diversion of ammonia to glutamate.

Several changes in tubular cell composition in addition to ammonia glutamate and α-ketoglutarate have been cited as possible factors controlling glutaminase activity. Hydrogen ion concentration seems improbable for maximum activity of the enzyme *in vitro* occurs at pH 9.0 and diminishes as pH decreases. One would anticipate that pH of the intracellular milieu of renal tubules would decrease in acidosis, from whatever its normal value may be, and thus reduce enzyme activity.

Depletion of body stores of potassium in man, dog and rat results in enhanced urinary excretion of ammonia (10, 22). Potassium depletion reduces ammonia excretion. Whether this is a consequence of induction of glutaminase by hypokalemia or of an increase in activity of a fixed amount of enzyme when cell potassium is reduced is debatable.

According to Preuss (19) the ratio of oxidized to reduced pyridine nucleotides controls the concentration of intracellular glutamate and thus controls glutaminase activity. Actually the role of these several variables has not been convincingly documented.

One possibility based on observations of Bendall and De Duve (3) seems to me to have been insufficiently explored. This is the concept that the permeability of mitochondria to glutamine may be enhanced in acidosis, hence permit entry of glutamine to the site of enzyme action. Actually the observations of Bendall and De Duve concerned the permeability of mitochondria of liver to glutamate. The extrapolation of mitochondria of kidney and glutamine may not be justified. However both glutaminase and glutamate dehydrogenase are intramitochondrial enzymes. If mitochondria were completely impermeable to glutamine in alkalosis, relatively impermeable in normal acid-base balance and quite permeable in acidosis one could explain most known facts concerning production and excretion of ammonia.

In conclusion: a number of factors affect the production of ammonia within renal tubular cells, hence its rate of excretion in the urine. It is doubtful that an adaptive increase of glutaminase as it occurs in the kidney of the rat, is a significant one. Neither is concentration, rate of filtration nor rate of reabsorption of the major precursor of ammonia, glutamine, a significant factor. What happens to glutamine, once it enters tubular cells, is significant. Also what happens to ammonia once it is produced within the cell is significant, for pathways of disposal alternate to excretion are available, including synthesis of glutamate and glutamine.

Various factors reputedly affect activity of glutaminase within tubular cells, including pNH_3, concentrations of α-ketoglutarate, glutamate, hydrogen ion, potassium ion and the ratio of oxidized to reduced pyridine nucleotides. Whether permeability of mitochondria to glutamine is a factor is uncertain.

It seems to me that there are two possible explanations of the adaptive increase in ammonia excretion in acidosis. First, a number of factors may increase ammonia excretion, no one of which by itself is an adequate quantitative explanation. The sum of their partial effects may be adequate. Second, if there is some unitary controlling factor, it has not yet been recognized. If one accepts any one of the presently proposed controlling factors as a complete and adequate explanation of ammonia production and excretion by the kidney, one does a disservice to science, for it discourages further inquiry.

SUMMARY

A variety of alterations of composition of extracellular fluid and of renal tubular cells have been cited as possible factors controlling production and excretion of ammonia by the kidney, no one of which constitutes an adequate quantitative explanation. Dog, rat and man all respond to an acid load with an adaptive increase in renal production and excretion of ammonia which reaches a maximum after 5 to 7 days, associated with a more or less severe acidosis. The rat responds with an adaptive increase in the activity of glutaminase-I, the major enzyme responsible for renal production of ammonia. The dog does not. Therefore enzyme adaptation is not obligatorily causal to increased excretion of ammonia. Neither is availability of substrate, glutamine, limiting. Rather, the activity of glutaminase within the mitochondria of tubular cells or the transport of glutamine to the active enzyme site is limiting. Within limits, the disposition of ammonia produced within the cell alternate to excretion is significant in the rat and guinea pig, not in the dog and cat. The several factors which may influence the intracellular activity of glutaminase *in vivo* include intracellular pH, intracellular K^+, glutamate, NH_3 and α-ketoglutarate concentrations and alterations of NAD/NADH ratio.

REFERENCES

1. Alleyene, G.A.O. and G.H. Scullard. *J. Clin. Invest.* 48:364, 1969.
2. Balagura-Baruch, S., L.M. Shurland and T. Welbourne. *Am. J. Physiol.* 218:1070, 1970.
3. Bendall, D.S. and C. De Duve. *Biochem. J.* 74:444, 1960.
4. Damian, A.C. and R.F. Pitts. *Am. J. Physiol.* 218:1249, 1970.
5. Davies, B.M.A. and J. Yudkin. *Biochem. J.* 52:407, 1952.
6. Denis, G., H. Preuss and R.F. Pitts. *J. Clin. Invest.* 43:571, 1964.

7. Goldstein, L. *Nature* 205:1130, 1965.
8. Goldstein, L. *Am. J. Physiol.* 213:983, 1967.
9. Goldstein, L. and J.M. Schooler. *Adv. Enz. Regulation* 5:71, 1967.
10. Huth, E.J., R.D. Squires and J.R. Elkinton. *J. Clin. Invest.* 38:1149, 1959.
11. Kamin, H. and P. Handler. *J. Biol. Chem.* 193:873, 1951.
12. Lyon, M.L. and R.F. Pitts. *Am. J. Physiol.* 216:117, 1969.
13. Oelert, H., E. Uhlich and A.G. Hills. *Pflüger's Arch. ges. Physiol.* 300:35, 1968.
14. Orloff, J. and R.W. Berliner. *J. Clin. Invest.* 35:223, 1956.
15. Pilkington, L.A., T.K. Young and R.F. Pitts. *Nephron* 7:51, 1970.
16. Pilkington, L.A., J. Welch and R.F. Pitts. *Am. J. Physiol.* 208:1100, 1965.
17. Pitts, R.F. *Fed. Proc.* 7:418, 1948.
18. Pitts, R.F. *Am. J. Physiol.* (submitted for publication).
19. Preuss, H.G. *J. Lab. Clin. Med.* 72:370, 1968.
20. Pollack, V.E., H. Mattenheimer, H. DeBruin and K.J. Weinman. *J. Clin. Invest.* 44:169, 1965.
21. Rector, F.C., Jr. and J. Orloff. *J. Clin. Invest.* 38:366, 1959.
22. Schwartz, W.B. and A.S. Relman. *J. Clin. Invest.* 32:258, 1953.

RELATION OF CARBOHYDRATE METABOLISM AND

AMMONIA PRODUCTION IN THE KIDNEY[1]

A. David Goodman,

*Division of Endocrinology and
Metabolism
Department of Medicine
Albany Medical College
Albany, New York   12208*

It is well established that in man, dog and rat the induction of metabolic acidosis leads to an increase in renal excretion of ammonia, which is due primarily to increased production of ammonia from glutamine (21, 26, 30, 31). The rise in ammoniagenesis from glutamine appears to be due in considerable measure to increased degradation of glutamine through a pathway involving deamidation of glutamine to glutamate by a mitochondrial phosphate-dependent glutaminase (PDG)[2] followed by deamination of glutamate to α-ketoglutarate (7, 8, 32). A second pathway, which probably is of less importance, consists of conversion to glutamine to α-ketoglutaramate by a specific glutamine

---

[1]This work was supported in part by U.S. Public Health Service Research Grant AM-09232.

[2]The abbreviations used are: PDG, phosphate dependent glutaminase and PEPCK, phosphoenolpyruvate carboxykinase.

transaminase, followed by deamidation of α-ketoglutaramate to α-ketoglutarate (8). The increase in metabolism of glutamine through the PDG pathway in acidosis cannot be ascribed to enhanced entry of this substrate into the cell since the renal concentration of glutamine is diminished in acidosis (8), and hence it seems likely that there is an increase in renal PDG activity. If, indeed, there is a rise in renal PDG activity in acidosis, in the dog it probably is due to activation of the enzyme *in vivo* rather than to increased enzyme synthesis (35). The same is probably true in the rat, for although there appears to be an increase in renal PDG synthesis in the acidotic rat, this increase is not indispensable to the rise in ammonia excretion in acidosis (10).

If acidosis directly causes activation of renal PDG it would be expected that the renal intracellular concentration of glutamate would be increased, since the PDG reaction is essentially unidirectional (17) and glutamate is a product of the reaction. Actually, the renal glutamate content is decreased in severe metabolic acidosis in both the rat and dog (9, 40). In a variety of biologic systems there exists a negative feedback relationship between the intracellular concentration of an immediate or distant product of an enzymatic reaction, and either the specific activity or the rate of synthesis of the enzyme (24). Goldstein has suggested that such a relationship may exist between glutamate and PDG, and that in acidosis the reduction in renal glutamate concentration causes activation of the enzyme (8, 9). Consistent with this hypothesis is the demonstration that glutamate, in the concentration in which it is found in renal cortical cells, is an inhibitor of solubilized PDG (9).

The possibility exists that the increase in generation of ammonia from glutamine in acidosis could be due primarily to an increased permeability of the mitochondria to glutamine, with a resultant increase in availability of glutamine to PDG. However, if the primary effect of acidosis were to enhance entry of glutamine into the mitochondria, one would expect the renal concentration of the product of the PDG reaction, glutamate, to be increased, whereas renal glutamate content is decreased in acidosis (9, 40).

Renal cortex is known to have a remarkable capacity to produce glucose from glutamate and its products (18) (Fig. 1), and hence we have investigated the possibility that acidosis may decrease cortical glutamate by accelerating its conversion to glucose. It was found that renal cortical slices taken from rats with metabolic acidosis have an increased capacity to convert glutamine and glutamate to glucose, and that the converse is true in alkalosis (11). Carbohydrate intake in the acidotic, control and alkalotic rats was the same for no solid food was provided, and equal amounts of glucose were fed by stomach tube to each group.[3] Acidosis stimulates production of glucose from α-ketoglutarate and oxalacetate, as well as from glutamine and glutamate, but does not increase glucose production from glycerol or fructose, suggesting that the stimulatory effect of acidosis on gluconeogenesis is probably due to increased

---

[3]Tube-feeding was facilitated by anesthetizing the rats with 50% CO_2 - 50% O_2 prior to intubation.

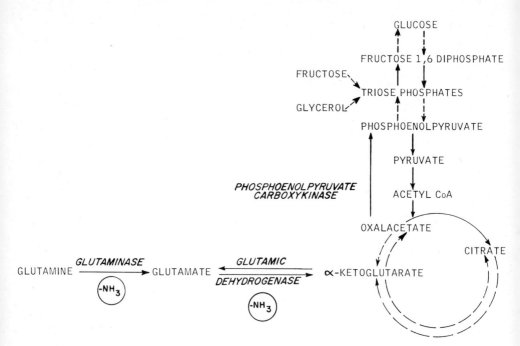

Fig. 1. Schematic diagram of the relation of renal cortical glutamine metabolism and gluconeogenesis.

activity of the rate-limiting gluconeogenic enzyme phosphoenolpyruvate carboxykinase (PEPCK) (11), and, indeed, it has been found that PEPCK activity in the cortex of animals in metabolic acidosis is substantially increased (1, 2, 36). It is of interest in this regard that in potassium deficiency, a state in which there is a considerable increase in ammonium ion excretion, renal cortex has an increased capacity to convert glutamine, glutamate, α-ketoglutarate and oxalacetate to glucose (11), and there is an increase in cortical PEPCK activity (16). Further, both cyclic adenosine monophosphate and cyclic inosine monophosphate, which increase the capacity of renal cortical slices to produce glucose from glutamine, glutamate, α-ketoglutarate and oxalacetate, but not from glycerol or fructose, cause a decrease in the glutamate content of cortical slices and a rise in production of ammonia from glutamine *in vitro* (28, 29).

On the basis of these findings we hypothesized that metabolic acidosis directly enhances conversion of glutamate and α-ketoglutarate to glucose by activation of PEPCK, and that the resultant decrease in intracellular glutamate or one of its products effects activation of PDG, thereby increasing ammonia production (11, 29). Implicit in this hypothesis are three basic assumptions: *one*, cortical glutamate or one of its products is a major determinant of PDG activity and ammonia production; *two*, the rate of gluconeogenesis affects substantially the cortical concentration of glutamate and its products; and *three*, renal gluconeogenesis *in vivo* is increased in metabolic acidosis. Recently, each of these assumptions has been examined experimentally.

Effect of renal cortical concentration of glutamate and its products on rate of ammoniagenesis:

The assumption that cortical glutamate itself is the major determinant of renal ammonia production is weakened by some of our recent observations. First, we have found that the administration of ammonia chloride to rats in an amount inadequate to decrease cortical glutamate (22 mmoles/kg/day for two days) increased the capacity of cortex to produce ammonia from glutamine, and that the administration of sodium bicarbonate (33 mmoles/kg/day for two days) decreased cortical production of ammonia without changing cortical glutamate (27). Second, we observed that cortex from potassium-depleted rats had an increased capacity to produce ammonia despite the fact that cortical glutamate content was normal (27). However, it is possible that in these experiments there occurred significant changes in the concentration of glutamate within the cortical mitochondrial compartment in which PDG is located, that were not reflected in total cortical glutamate concentration. Further, even if one cannot ascribe the increase in ammonia production in metabolic acidosis and potassium depletion to a decrease in cortical glutamate, the fact that cortical glutamate does not increase in these states despite a presumed increase in the generation of glutamate from glutamine is of importance, for a rise in glutamate would tend to decrease net production of ammonia in the rat by enhancing resynthesis of glutamine through the glutamine synthetase pathway (7). Thus, even if accelerated removal of the products of glutamine degradation may not be the primary cause of the increase in production of ammonia in acidosis, it may contribute to maintaining increased ammoniagenesis.

In a related study, we have incubated cortical slices from normal animals in modified Krebs-Ringer bicarbonate containing 10 mM glutamine at pH 7.1 (12 mEq/l bicarbonate), 7.4 (24 mEq/l bicarbonate), and 7.7 (48 mEq/l bicarbonate) and observed that although the glutamate content of the slices at pH 7.1 was significantly decreased, and at pH 7.7 it was significantly increased, ammonia production from glutamine was not affected by the alteration of medium pH (27). These findings suggest that glutamate content is not a major determinant of ammonia production at least *in vitro*. However, they do not exclude the possibility that in metabolic acidosis the observed fall in renal concentration of α-ketoglutarate (1, 9, 33) or a change in the concentration of one of its products may be the cause of the increase in PDG activity and ammoniagenesis, for whereas in acidosis *in vivo* the renal content of α-ketoglutarate decreases, in rat cortical slices incubated in an acidic medium *in vitro* this does not occur (33).

Another point that casts doubt on the concept that glutamate concentration regulates PDG activity is the observation that intravenous infusion of glutamic acid in the dog does not decrease urinary excretion of ammonium (22). One must consider the possibility, however, that exogenously administered glutamate may not penetrate readily into the mitochondria, where PDG is located (32).

Balagura-Baruch et al. have observed that intravenous infusion of α-ketoglutaric acid in the dog causes a considerable decrease in renal ammonia production (3). Renal glutamate production was determined during the infusion and it was found that only 25% of the decrease in ammonia production could be explained by condensation of ammonia with α-ketoglutarate to form glutamate, while condensation of ammonia with glutamate to form glutamine was excluded by the fact that dog kidney lacks glutamine synthetase (23). Therefore, the suppression of renal ammonia production by intravenous infusion of α-ketoglutarate suggests strongly that either α-ketoglutarate itself, its reductive product glutamate, or one of its oxidative products such as succinate or fumarate, inhibits PDG. Consistent with this conclusion is the observation that infusion of α-ketoglutaric acid in the rat decreases glutaminase activity as measured *in vivo* (7).

These observations lend credence to the hypothesis that in acidosis the decrease in cortical concentration of glutamate, α-ketoglutarate or the oxidative products of α-ketoglutarate contributes to the increase in production of ammonia from glutamine.

Effect of gluconeogenesis on renal cortical glutamate and α-ketoglutarate content and on ammonia production:

The hypothesis that the decrease in the renal content of glutamate and α-ketoglutarate in acidosis is due to accelerated conversion of these compounds to glucose, is based on the assumption that in the renal cortex gluconeogenesis constitutes a quantitatively important pathway for the disposition of the carbon skeleton of glutamine following its deamidation and deamination. Kamm and Asher have examined this problem in an *in vitro* system in which they measured the simultaneous production of glucose and of ammonia by renal cortical slices incubated in a glucose-free medium containing 10 mM glutamine, and they have deduced that following deamidation and deamination about 75% of the carbon skeleton of glutamine is ultimately converted to glucose (13). Further, in a similar study using glutamate they concluded that about 93% of the glutamate that is deaminated by cortex *in vitro* is converted to glucose (13). Of course, one cannot be certain that the same holds true *in vivo*.

Churchill and Malvin recently have challenged the hypothesis that in metabolic acidosis the increase in renal ammonia production is mediated by an increase in renal gluconeogenesis (6). They base their criticism in part on the observation that intravenous infusion of sodium L-lactate in the dog (1.7 mmoles/min) increases renal production of glucose *in vivo*, as determined by measurement of arterial and renal venous glucose concentration, but decreases ammonium excretion. In interpreting this study, it should be considered that the increase in gluconeogenic capacity in metabolic acidosis and the increase in gluconeogenesis during lactate infusion are probably mediated by entirely different mechanisms. The increase in gluconeogenic capacity in acidosis is due presumably to accelerated conversion of oxalacetate to phosphoenolpyruvate, which would decrease the concentration of gluconeogenic substrates antecedent to this reaction, such as glutamate and α-ketoglutarate, and thereby presumably activate PDG. In contrast, the increase in gluconeogenesis effected by infusion of lactate probably is due to the fact that lactate is itself a gluconeogenic substrate (18), and thus there would be no reason to expect that lactate infusion would cause an increase in ammoniagenesis. It should be emphasized that according to the "PEPCK-ammoniagenesis hypothesis" both the increase in gluconeogenic capacity and the increase in ammoniagenesis are secondary to increased PEPCK activity, and the hypothesis does not imply that increases in gluconeogenesis due to causes other than activation of PEPCK should stimulate ammonia production. Incidentally, the decrease in ammonium excretion following infusion of sodium L-lactate could be due to the fact that lactate is metabolized to bicarbonate (12).

In vitro demonstrations of a dissociation between gluconeogenesis and ammoniagenesis have also been made. Working with renal cortical slices, we have found that either omission of calcium from the incubation medium or addition of theophylline to the medium markedly decreases cortical production of glucose from glutamine and glutamate, without decreasing ammonia production (27), and others have observed that phenylpyruvate suppresses cortical gluconeogenesis *in vitro* but stimulates ammoniagenesis (5). However, in our studies the omission of calcium and addition of theophylline did not increase cortical glutamate concentration, in contrast with the situation in severe alkalosis (9), and in the phenylpyruvate experiment the glutamate concentration was not measured. The absence of a rise in cortical glutamate in the calcium and theophylline experiments suggests that although conversion of glutamate to glucose was impaired there may have been an increase of equal or greater magnitude in the conversion of glutamate to CO_2 through the tricarboxylic acid cycle or the hexose monophosphate shunt. Indeed, the decrease in gluconeogenesis from glutamate in these experiments could have been secondary to increased oxidation of glutamate to CO_2 with resultant diversion of glutamate from the gluconeogenic pathway.

It has been observed by Churchill and Malvin that when sliced cortex from acidotic dogs is incubated in a phosphate buffer its capacity to produce ammonia from glutamine is not increased, as compared to cortex from normal dogs, despite the fact that its gluconeogenic capacity is increased (6). This observation has been interpreted as ruling against a causal relationship between the increase in gluconeogenesis and the increase in ammoniagenesis occurring in acidosis *in vivo*. It is not clear why in this experiment there was no increase in ammonia production by the slices from the acidotic dogs, since renal production of ammonia is definitely increased in metabolic acidosis *in vivo* in the dog (31), and it has been demonstrated by several investigators that renal cortical slices from acidotic rats have an increased capacity to produce ammonia from glutamine when incubated in bicarbonate buffer (1, 13, 27). Perhaps the use of phosphate-buffered saline as the incubation medium in the experiment of Churchill and Malvin may account for the absence of an increase in ammoniagenesis in their study.

Preuss has demonstrated an apparent dissociation between gluconeogenesis and ammonia production *in vitro* in an experiment using malonate. He observed that cortical slices incubated at pH 7.1 produce more glucose and ammonia from glutamate than slices incubated at pH 7.7, and that when malonate (20 mM) is added to both media, ammonia production continues to be higher in the pH 7.1 medium despite the fact that glucose production in both media is almost completely abolished (33). The physiologic significance of this study is uncertain. First, as indicated by Kamm and Asher (13), 20 mM malonate effects only a partial inhibition of the conversion of α-ketoglutarate to succinate, and the marked suppression of gluconeogenesis caused by this agent could be due in part to the fact that it inhibits reactions in the gluconeogenic pathway beyond the PEPCK reaction; consequently, increased PEPCK activity in the presence of these inhibitors might still produce considerable acceleration of the conversion of glutamate to phosphoenolpyruvate, and the latter ultimately could be oxidized to CO_2 in the tricarboxylic acid cycle. Second, the mechanism by which incubation of cortical slices *in vitro* at low pH causes an increase in production of ammonia from glutamate may be quite different than the mechanism by which chronic acidosis *in vivo* causes a rise in the capacity of cortex to produce ammonia from glutamine. In support of this contention it is of note that incubation of cortical slices at low pH does not increase production of ammonia from glutamine, whereas cortex from acidotic rats has an increased capacity to produce ammonia from glutamine (27); that incubation at low pH increases the α-ketoglutarate content of rat cortical slices (33), whereas renal α-ketoglutarate is decreased substantially in acidotic rats (1, 9, 33); and that incubation at low pH increases production of CO_2 from glutamine both in slices and mitochondria, whereas only slices but not mitochondria from acidotic dogs have an increased capacity to produce CO_2 from glutamine (39).

Effect of metabolic acidosis on renal gluconeogenesis in vivo:

Despite the fact that renal cortical slices and the perfused kidney preparation have a remarkable capacity to produce glucose from a number of substrates *in vitro* (4, 18), studies in normal, intact dogs suggest that there is little if any net renal production of glucose *in vivo*. Specifically, in several recent studies in anesthetized dogs fasted for 18 to 24 hr, no significant difference was observed between the glucose concentration in renal venous blood and in arterial blood (6, 37, 40). However, the renal venous-arterial difference for glucose *in vivo* is affected by the metabolism of medulla as well as of cortex. Hence, it is possible that cortical production of glucose *in vivo* is substantial but that this is offset by utilization of glucose by medulla, which is known to depend primarily on glucose for its energy (20).

Churchill and Malvin were not able to detect a significant change in renal glucose output *in vivo* in dogs made acidotic by feeding of NH_4Cl for five days (6), and Roxe, DiSalvo and Balagura-Baruch did not observe an increase in renal glucose output in dogs in whom acute metabolic acidosis was induced by intravenous infusion of sulfuric acid (37). In both studies, a conventional manual technique for measuring blood glucose was employed. Steiner, Goodman and Treble, using a highly precise, automated method for measuring blood glucose, found an extremely small but statistically significant increase in renal glucose output *in vivo* in dogs with metabolic acidosis of three days duration; the mean venous-arterial difference (corrected for extraction of urine) in 19 control animals was 0.09 ± 0.10 (S.E.) mg/100 ml, and in 16 acidotic dogs it was 0.37 ± 0.07 ($p < 0.05$). while the renal blood flow in the two groups was almost identical (40). Despite the fact that the difference between the two groups was statistically significant, it was so very small that the study cannot be taken as conclusive proof of an effect of acidosis on renal gluconeogenesis *in vivo*; minor unrecognized artefacts in the measurement technique or small differences in red cell or medullary glucose utilization in the two groups could readily account for the minute difference observed between the two groups.

The fact that there is no easily detectable increase in renal glucose output *in vivo* in acidosis does not rule against the hypothesis that in metabolic acidosis there is a substantial increase in the conversion of glutamine to glucose, for the blood flow of the kidney is so high that even if all of the increment in glutamine being extracted by the kidney in acidosis was converted to glucose, the resultant increase in renal venous glucose would be exceedingly small. As an illustration, in the study of Shalhoub *et al.* the renal arterial-venous plasma difference for glutamine in a group of acidotic dogs exceeded that of a group of alkalotic dogs by 0.08 μmoles/ml (38); if all of the 0.08 μmoles/ml of glutamine were converted to glucose, renal venous plasma would increase by only 0.04 μmoles/ml[4] or 0.7 mg% -- a change that would be difficult to detect even with replicate analyses. Moreover, if there is an increase in medullary glucose utilization in acidosis, the venous-arterial glucose difference would be even smaller.

Actually, it would not be surprising if in the future it is demonstrated definitively that cortical gluconeogenesis is not increased *in vivo* in acidosis. In *in vitro* studies when cortical slices are incubated in media devoid of glucose (11, 15), the increase in cortical PEPCK activity in acidosis appears to cause an increase in the conversion of glutamine and its products to glucose, but this may not be the case *in vivo* where there is considerable glucose in the extracellular fluid and the concentration of cortical intracellular metabolites may be quite different than *in vitro*. Nevertheless, the

---

[4] In the course of conversion of glutamine to glucose, 2 μmoles of glutamine are converted to 2 μmoles of triose which are then condensed to form 1 μmole of glucose.

increase in cortical PEPCK activity *in vivo* might enhance removal of the products of glutamine degradation by accelerating their oxidation to CO_2, either through the pathway involving conversion of oxalacetate to phosphoenolpyruvate, transformation to pyruvate and then acetate, and oxidation of acetate in the tricarboxylic acid cycle (Fig. 1), or through the hexose monophosphate shunt. In this regard it should be emphasized that α-ketoglutarate must be converted to phosphoenolpyruvate or pyruvate in order to be oxidized completely in the tricarboxylic acid cycle. This is so because the compounds of the tricarboxylic acid cycle cannot be dissipated solely through the reactions of the cycle *per se*, for although with each passage of an intermediary compound around the cycle two carbon atoms are removed as CO_2, two additional carbon atoms are added as acetate. Consequently, α-ketoglutarate can be oxidized to CO_2 only if the oxalacetate arising from it "escapes" the cycle by conversion to phosphoenolpyruvate under the influence of PEPCK, or by conversion to pyruvate, and then reenters the cycle as acetate. Thus, PEPCK activity may be rate-limiting not only for cortical conversion of α-ketoglutarate to glucose but also for oxidation of α-ketoglutarate to CO_2. In view of these facts one should qualify the initial "PEPCK ammoniagenesis hypothesis" by stating that the increase in cortical PEPCK activity in acidosis may enhance removal of the carbon skeleton of glutamine *in vivo* by accelerating its oxidation to carbon dioxide as well as by enhancing its conversion to glucose.

Simpson and Sherrard have found that kidney slices from acidotic dogs have an increased capacity to produce $^{14}CO_2$ from glutamine-U-^{14}C

(39), but this does not necessarily mean that in acidosis there is an increase in total oxidation of glutamine to CO_2 in the tricarboxylic acid cycle. The rise in $^{14}CO_2$ production could be secondary to increased conversion of the glutamine-U-^{14}C to glucose, for as glutamine is converted to glucose two molecules of CO_2 are generated from each molecule of glutamine in the portion of the gluconeogenic pathway between α-ketoglutarate and phosphoenolpyruvate. Interestingly, in contrast with kidney slices from acidotic dogs, isolated mitochondria from acidotic dogs do not have an increased capacity to produce $^{14}CO_2$ from glutamine-U-^{14}C, suggesting that the increase in CO_2 generation in the slices may be due to acceleration of some extramitochondrial reaction (39). PEPCK is located in the cytoplasm in the rat, but it is restricted to the mitochondria in rabbit and chicken (19) and its intracellular location in the dog has not been ascertained. If the PEPCK of dog kidney proves to be cytoplasmic, one might reasonably hypothesize that the increase in CO_2 production from glutamine that is observed in kidney slices but not kidney mitochondria from acidotic dogs is due to increased PEPCK activity.

It has been demonstrated by Preuss that in metabolic acidosis in the rat there is an increase in cortical NAD and in the NAD/NADH$_2$ ratio (34), and he has made the interesting suggestion that since NAD is a cofactor in the conversion of glutamate to α-ketoglutarate, α-ketoglutarate to succinate and malate to oxalacetate, the increase in NAD would accelerate removal of glutamate and thereby enhance ammoniagenesis. Consistent with this hypothesis is the observation that administration to rats of nicotinamide, which increases cortical

NAD and $NAD/NADH_2$, decreases cortical glutamate and increases renal ammonia excretion. It should be emphasized the Preuss's "NAD-ammoniagenesis hypothesis" and the "PEPCK-ammoniagenesis hypothesis" are not mutually exclusive, and that in acidosis the increase in $NAD/NADH_2$ and the increase in PEPCK activity in combination may produce a greater acceleration in conversion of glutamate to glucose and/or CO_2 than would be effected by either factor alone. It should also be noted that an increase in $NAD/NADH_2$ alone would be expected to increase the renal concentration of α-ketoglutarate (1, 9, 33). One can reasonably explain the fall in α-ketoglutarate on the basis of the increase in PEPCK activity that occurs in combination with the rise in $NAD/NADH_2$.

It is reasonable to assume that the rise in cortical $NAD/NADH_2$ in acidosis is due to an increase in this ratio in the mitochondria rather than in the cytoplasm, since the pyruvate/lactate ratio, which is thought to reflect cytoplasmic $NAD/NADH_2$, is not changed in acidosis (1). Possibly the increase in PEPCK activity in acidosis is the cause of the increase in $NAD/NADH_2$ through the following sequence of events: increased PEPCK would accelerate conversion of oxalacetate to phosphoenolpyruvate, thereby decreasing the amount of oxalacetate available to condense with acetate, reducing the rate of oxidation of acetate in the tricarboxylic acid cycle, and slowing the conversion of NAD to $NADH_2$.

The increase in renal ammonia production that occurs in metabolic acidosis is a physiologic phenomenon that is of importance to survival, and it is becoming increasingly apparent that critical homeostatic systems are usually regulated by more than one mechanism. Although in this discussion we have emphasized the potential significance of enhanced removal of glutamate and its products in the acceleration of ammoniagenesis in acidosis, it seems very likely that other factors play an important role. Indeed, it has already been demonstrated that a decrease in glutamine synthetase activity contributes to the increase in ammonia production in the acidotic rat (7). It is also possible that in acidosis there is activation of PDG and glutamine transaminase by factors other than alteration in the concentration of their end products, and possibly increased mitochondrial permeability to glutamine may play a contributory role.

Renal gluconeogenesis and ammoniagenesis in prolonged starvation:

Owen *et al*. have found that in obese humans fasted for five to six weeks renal glucose production is substantial and indeed is almost equal to hepatic glucose production. In determining renal venous-arterial glucose difference in this study, the venous glucose concentration was not corrected for the decrease in plasma volume that occurs across the kidney due to extraction of urine, but the urine volume in these patients was very low (personal communication). The difficulties inherent in measuring accurately small renal venous-arterial glucose differences must be borne in mind, but nevertheless these observations are of considerable interest. It is relevant that renal cortex taken from rats deprived of food for several days has an increased capacity to produce glucose *in vitro*, and the increase in gluconeogenic capacity appears to be secondary in large measure to the metabolic acidosis that occurs as a consequence of starvation ketosis (14), suggesting that in starved human beings the relatively high rate of glucose production may be secondary to the existing ketoacidosis.

Urinary ammonium ion excretion is increased in prolonged starvation, presumably secondary to the associated acidosis, and urea excretion is markedly decreased (25). The relatively high rate of renal gluconeogenesis may reflect renal production of glucose from the carbon skeleton of amino acids following generation of ammonia from their amide and mine groups. Owen *et al.* have suggested that renal glucose production may play a role in conservation of protein in starvation in that the high rate of renal gluconeogenesis would tend to decrease the requirement for hepatic gluconeogenesis, thereby minimizing hepatic urea production and total nitrogen excretion.

SUMMARY

In metabolic acidosis there is increased renal cortical phosphoenolpyruvate carboxykinase (PEPCK) activity, and this probably accounts for the increased capacity of cortical slices from acidotic rats to convert glutamine, glutamate, α-ketoglutarate and other substrates to glucose. The decrease in renal glutamate and α-ketoglutarate concentration in acidosis presumably is due at least in part to the enhanced conversion of these compounds to glucose. It is postulated that glutamate, α-ketoglutarate or one of their derivatives is an "end-product inhibitor" of phosphate-dependent glutaminase, and that the decrease in concentration of this inhibitor causes activation of glutaminase and thereby contributes to the rise in ammoniagenesis from glutamine in acidosis. The increased PEPCK activity may also accelerate removal of the products of glutamine degradation by enhancing their oxidation to CO_2. These hypotheses do not exclude the possibility that mechanisms unrelated to PEPCK activity are important in the regulation of ammoniagenesis. The question of whether cortical gluconeogenesis *in vivo* is increased in metabolic acidosis remains unresolved.

In prolonged starvation with secondary ketoacidosis, renal conversion of amino acids to glucose, following ammoniagenesis from their amide and amino groups, may decrease the requirement for hepatic gluconeogenesis and thus minimize protein catabolism.

REFERENCES

1. Alleyne, G.A. *J. Clin. Invest. 49*:943, 1970.
2. Alleyne, G.A. and G.H. Scullard. *J. Clin. Invest. 48*:364, 1969.
3. Balagura-Baruch, S., L.M. Shurland and T.C. Welbourne. *Amer. J. Physiol. 218*:1065, 1970.
4. Bowman, R.H. *J. Biol. Chem. 245*:1604, 1970.
5. Churchill, P.C. and R.L. Malvin. *Amer. J. Physiol. 218*:35, 1970.
6. Churchill, P.C. and R.L. Malvin. *Amer. J. Physiol. 218*:241, 1970.
7. Damian, A.C. and R.F. Pitts. *Amer. J. Physiol. 218*:1249, 1970.
8. Goldstein, L. *Amer. J. Physiol. 213*:983, 1968.
9. Goldstein, L. *Amer. J. Physiol. 210*:661, 1966.
10. Goldstein, L. *Nature 205*:1330, 1965.
11. Goodman, A.D., R.E. Fuisz and G.F. Cahill, Jr. *J. Clin. Invest. 45*:612, 1966.

12. Hartmann, A.F. and M.J. Senn. *J. Clin. Invest.* *11*:327, 1932.
13. Kamm, D.E. and R.R. Asher. *Amer. J. Physiol.* *218*:1161, 1970.
14. Kamm, D.E. and G.F. Cahill. *Amer. J. Physiol.* *216*:1207, 1969.
15. Kamm, D.E., R.E. Fuisz, A.D. Goodman and G.F. Cahill, Jr. *J. Clin. Invest.* *46*:1172, 1967.
16. Kamm, D. and G. Strope. *Clin. Res.* *18*:504, 1970.
17. Klingman, J.D. and P. Handler. *J. Biol. Chem.* *232*:369, 1958.
18. Krebs, H., D. Bennett, P. De Gasquet, T. Gascoyne and T. Yoshida. *Biochem. J.* *86*:22, 1963.
19. Lardy, H.A. *Harvey Lectures* *60*:261, 1964-65.
20. Lee, J.B., V.K. Vance and G.F. Cahill, Jr. *Amer. J. Physiol.* *203*:27, 1962.
21. Leonard, E. and J. Orloff. *Amer. J. Physiol.* *182*
22. Lotspeich, W.D. and R.F. Pitts. *J. Biol. Chem.* *168*:611, 1947.
23. Lyon, M.C. and R.F. Pitts. *Amer. J. Physiol.* *216*:117, 1969.
24. Moyed, H.S. and H.E. Umbarger. *Physiol. Rev.* *42*:444, 1962.
25. Owen, W.E., P. Felig, A.P. Morgan, J. Wahren and G.F. Cahill, Jr. *J. Clin. Invest.* *48*:574, 1969.
26. Owen, E. and R. Robinson. *J. Clin. Invest.* *42*:263, 1963.
27. Pagliara, A.S. and A.D. Goodman. *J. Clin. Invest.* (in press).
28. Pagliara, A.S. and A.D. Goodman. *Amer. J. Physiol.* *218*:1301, 1970.
29. Pagliara, A.S. and A.D. Goodman. *J. Clin. Invest.* *48*:1408, 1969.
30. Pitts, R.F. *Amer. J. Med.* *36*:720, 1964.
31. Pitts, R.F., J. De Haas and J. Klein. *Amer. J. Physiol.* *204*:187, 1963.

32. Pitts, R.F., L.A. Pilkington and J.C.M. De Haas. *J. Clin. Invest.* 44:731, 1965.
33. Preuss, H.G. *Nephron* 6:235, 1969.
34. Preuss, H.G. *J. Clin. Lab. Med.* 72:370, 1968.
35. Rector, F.C., Jr. and J. Orloff. *J. Clin. Invest.* 38:366, 1959.
36. Rosenzweig, S. and D. Frascella. *Bull. New Jersey Acad. Sci.* 13:17, 1968.
37. Roxe, D.M., J. Di Salvo, and S. Balagura-Baruch. *Amer. J. Physiol.* 218:1676, 1970.
38. Shalhoub, R., W. Webber, S. Glabman, M. Canessa-Fischer, J. Klein J. De Haas and R.F. Pitts. *Amer. J. Physiol.* 204:181, 1963.
39. Simpson, D.P. and D.J. Sherrard. *J. Clin. Invest.* 48:1008, 1969.
40. Steiner, A.L., A.D. Goodman and D.H. Treble. *Amer. J. Physiol.* 215:211, 1968.